IN MEMORY OF

MILLIS, MASSACHUSETTS
BENNETT GENERAZIO

W9-BQZ-672

523
UNIV

523
UNIV   Universe explained

A 5439

FEB 21   H 4464
MAR 8   A 6048
JUN 2   A 2773
AUG 7   A 5881
FEB 21

Millis Public Library
Auburn Road
Millis, Mass. 02054

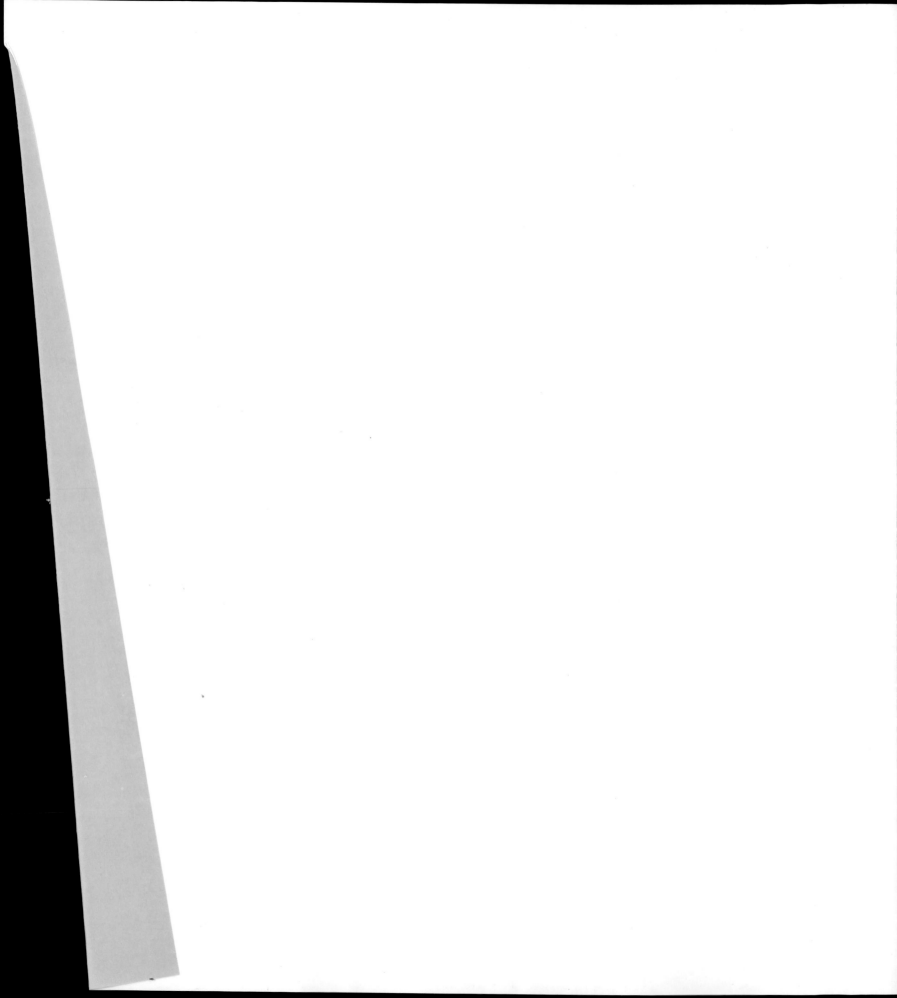

The UNIVERSE EXPLAINED

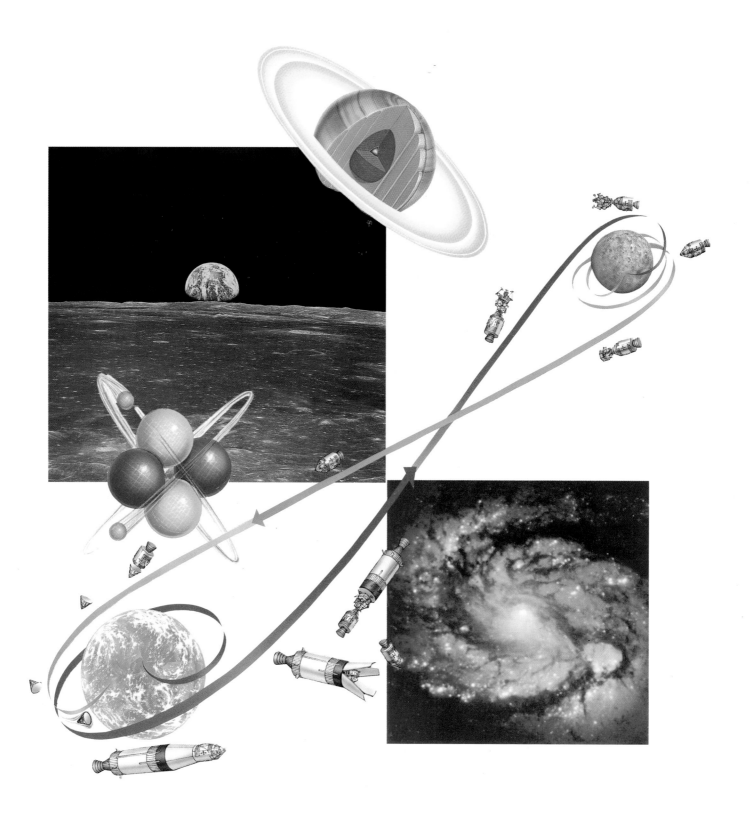

Millis Public Library
Auburn Road
Millis, Mass. 02054

DEC 1 0 1994

The UNIVERSE EXPLAINED

The Earth-Dweller's Guide to the Mysteries of Space

Colin A. Ronan

A HENRY HOLT REFERENCE BOOK

HENRY HOLT AND COMPANY

NEW YORK

Contents

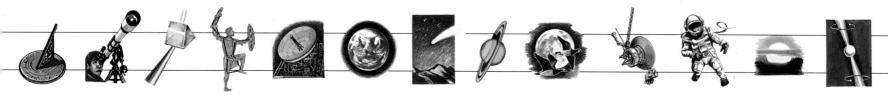

A Henry Holt Reference Book
Henry Holt and Company, Inc.
Publishers since 1866
115 West 18th Street
New York, New York 10011

Henry Holt ® is a
registered trademark of
Henry Holt and Company, Inc.

Copyright © 1994 by
Marshall Editions Developments Limited
All rights reserved.

**Library of Congress
Cataloging-in-Publication Data**

The Universe explained : the Earth-dweller's
 guide to the mysteries of space / Colin A. Ronan,
 general editor. — 1st ed.
   p. cm. — (Henry Holt reference book)
 Includes index.
 1. Astronomy—Popular works.
 2. Cosmology—Popular works.
 I. Ronan, Colin A. II. Series.
 QB44.2.U55   1994
 523—dc20   94-16294
                                  CIP

 ISBN 0-8050-3488-9

Henry Holt Books are available for special
promotions and premiums. For details
contact: Director, Special Markets.

First Edition—1994

Conceived, edited, and designed by
Marshall Editions, London
Printed and bound in Italy by
New Interlitho Italia, Milan
Origination by HBM Print Pte, Singapore
Filmsetting supplied by
Dorchester Typesetting Group Limited

All first editions are printed on acid-free paper ∞

10 9 8 7 6 5 4 3 2 1

Project editor   Jon Kirkwood
Art editor   Simon Adamczewski
Assistant editor   Jon Richards
Picture research   Vanessa Fletcher

Contributors   Nigel Cawthorne
    John Farndon
    Robin Scagell

*Previous page (clockwise from top):
cross-section of the gas giant Saturn; lord
of the rings; exploring the Moon with the
Apollo missions; a spiral galaxy; matter's
building blocks; planet Earth rises over the
barren moonscape.*

*Overleaf (clockwise from top):
the island in space that is a galaxy;
viewing the sky through binoculars;
poisonous gas clouds swirl above Venus;
darkness in the shadow of a hand; a satellite
launched to measure high-energy
radiation.*

Foreword

The starry heavens existed long before life began on Earth. When our human ancestors first walked the African plains, the sky already formed a canopy over us, and the study of the Sun, Moon, and stars proved an unending source of fascination. Over the ages, musicians have sung its praises, and painters and poets still record its glories.

But now in the late 20th century, we can probe space in ways never before possible to the human race. We possess telescopes of unprecedented power, which we can use not only to see stars and galaxies, but also to detect the radio waves and heat rays they emit. Then, by launching satellites into orbit around the Earth, we can observe the normally invisible X-rays and other forms of short-wave radiation from outer space. Such satellites have also made it possible to study the planets and their moons, as well as comets and the other debris of the solar system, in ways which no astronomer ever dreamed of when the century dawned.

This book describes the quite astounding results of all these observations and explains the exciting theories that have grown out of them. We explore the planets; we examine the Sun and the stars, analyze the principles which make them shine, and then find out how they are born and how they die. We also investigate the very depths of space to discover the universe of galaxies, where vast collections of stars, dust, and gas are undergoing giant explosions and collisions. We even consider how the whole universe began and how it may end. Altogether, it makes a fascinating story of awe-inspiring dimensions.

Colin A. Ronan

Introduction

Across five interlinked sections, each probing a different aspect of the cosmos, this book surveys our knowledge of the universe, considers the theories of its evolution, and examines its workings. In this way the explanations of how the pieces of the jigsaw puzzle fit together become clear and make this most exciting field of study accessible and understandable.

OBSERVATORY EARTH

Whether watching the Moon's phases or picking up faint radio signals, **Observatory Earth** shows how astronomy makes sense of the data pouring in from space to Earth and sets a scale for measuring the universe.

THE PLANETS

There are similarities between the bodies that orbit the Sun, but also many differences. **The Planets** surveys the evidence, maps the features and probes to the core of each planet, then explores to the remote edges of the solar system.

SUN AND STARS

Our sun is a fairly ordinary, stable star. **Sun and Stars** explains why it and all the other stars are as they are, from giant, short-lived blue stars to collapsed stellar remnants where matter is crushed out of existence.

NEBULAE AND GALAXIES

Vast clouds and banks of stars make up galaxies, star cities that can have the most delicate spiral structures. **Nebulae and Galaxies** looks from inside our own galaxy out to the largest structures known – superclusters of galaxies.

HOW THE UNIVERSE WORKS

From the forces that govern every single event to the events that have formed the universe as we know it, **How the Universe Works** steadily uncovers and simplifies the layers of complexity at the frontiers of astronomical thought and theory, clarifying them with everyday examples.

It is relatively simple merely to describe the wonders of the universe, but much more rewarding to understand why the universe behaves as it does. Using a wealth of analogies with familiar items and events, and with the help of straightforward language, this book provides explanations of a kind not otherwise easily attainable.

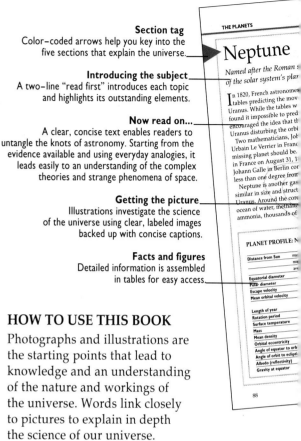

Section tag
Color–coded arrows help you key into the five sections that explain the universe.

Introducing the subject
A two–line "read first" introduces each topic and highlights its outstanding elements.

Now read on...
A clear, concise text enables readers to untangle the knots of astronomy. Starting from the evidence available and using everyday analogies, it leads easily to an understanding of the complex theories and strange phenomena of space.

Getting the picture
Illustrations investigate the science of the universe using clear, labeled images backed up with concise captions.

Facts and figures
Detailed information is assembled in tables for easy access.

HOW TO USE THIS BOOK

Photographs and illustrations are the starting points that lead to knowledge and an understanding of the nature and workings of the universe. Words link closely to pictures to explain in depth the science of our universe.

Each of the book's double-page spreads is a self-contained story. However, explaining the phenomena of astronomy is complex, and topics do not always fit neatly and completely under the headings superimposed on them. To deal with this, connections lead from the edge of each right-hand page to other topics, both within the same section and in different sections.

In order to promote flexibility of thought and depth of interest, the connections made are often deliberately wide-ranging. Using the connections will make the book fully interactive and forge the links of understanding between the different, yet complementary, aspects of astronomy.

The three sections of the book devoted to specific types of heavenly objects – **The Planets**, **Sun and Stars**, and **Nebulae and Galaxies** – not only describe but also explain the phenomena they deal with in detail. **Observatory Earth** and **How the Universe Works** explore the practice and theory of astronomy and provide a valuable resource for deeper understanding.

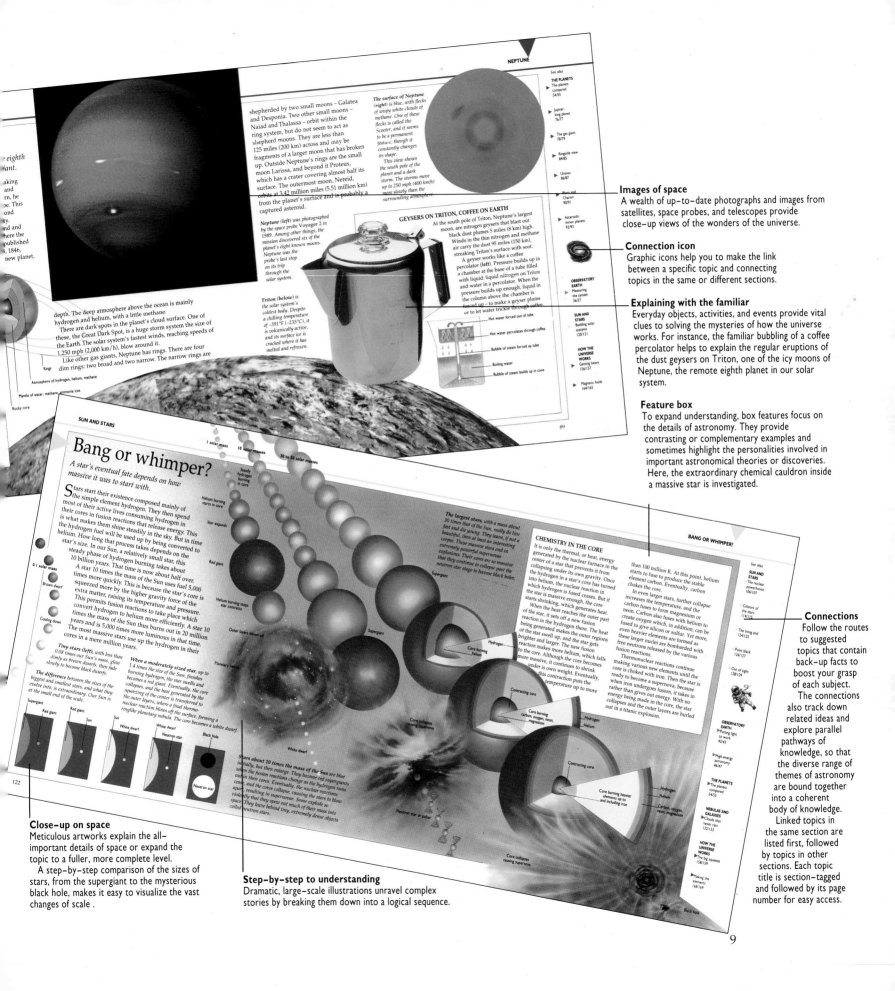

Images of space
A wealth of up-to-date photographs and images from satellites, space probes, and telescopes provide close-up views of the wonders of the universe.

Connection icon
Graphic icons help you to make the link between a specific topic and connecting topics in the same or different sections.

Explaining with the familiar
Everyday objects, activities, and events provide vital clues to solving the mysteries of how the universe works. For instance, the familiar bubbling of a coffee percolator helps to explain the regular eruptions of the dust geysers on Triton, one of the icy moons of Neptune, the remote eighth planet in our solar system.

Feature box
To expand understanding, box features focus on the details of astronomy. They provide contrasting or complementary examples and sometimes highlight the personalities involved in important astronomical theories or discoveries. Here, the extraordinary chemical cauldron inside a massive star is investigated.

Connections
Follow the routes to suggested topics that contain back-up facts to boost your grasp of each subject. The connections also track down related ideas and explore parallel pathways of knowledge, so that the diverse range of themes of astronomy are bound together into a coherent body of knowledge. Linked topics in the same section are listed first, followed by topics in other sections. Each topic title is section-tagged and followed by its page number for easy access.

Close-up on space
Meticulous artworks explain the all-important details of space or expand the topic to a fuller, more complete level.
A step-by-step comparison of the sizes of stars, from the supergiant to the mysterious black hole, makes it easy to visualize the vast changes of scale.

Step-by-step to understanding
Dramatic, large-scale illustrations unravel complex stories by breaking them down into a logical sequence.

Observatory Earth

Astronomy, like other sciences, is about seeking the answers to questions. Why does the Sun rise and set? Why does the Moon have phases? How distant are planets and stars? How big is the universe? How did it begin – and how may it end?

Theories can attempt to answer these questions, but only by observation can the hard evidence needed to prove or disprove a theory be attained. From Earth, our own "space station," we look out into the universe, and by observing the way our Earth moves through space, we have learned how it fits into the scheme of things. And gradually astronomers have made sense of the data that streams in to us from the universe. By analysis and interpretation, they attempt to answer some of the big questions about the universe and the way it works.

Left (clockwise from top): the Moon's phases; gamma rays; early star map; radio astronomy; X-rays; using a telescope.
This page (top): rays from the Sun; (left) the spectrum of colors.

In a spin

We now know that the Earth spins once a day as it orbits the Sun. But what evidence is there to demonstrate that this is the case?

From an Earth-dweller's viewpoint, it would seem obvious to believe that the Sun and Moon and all the other celestial bodies move around the Earth once a day. For they all rise in the eastern part of the sky and set in the west, while the Earth itself is clearly immobile. Although some ancient Greek astronomers had suggested that the movements of the heavens could just as well be accounted for by a moving Earth, it seemed simpler to assume that the Earth remained still with everything else moving around it.

It was a Polish scholar, Copernicus, who investigated how this Earth-centered system could give predictions of the positions of the planets. In 1543 he published a book in which he set out how the movements of the planets could be accounted for just as well if the Sun, rather than the Earth, was at the center of the solar system.

A key point of the theory was that the Earth was a globe that rotated on its axis. Critics pointed out that if the Earth rotated, a ball thrown upward should actually land slightly to one side because the Earth was moving beneath it. We now know that effects like this do occur, but they are subtle and can be measured only using precise instruments not available to scientists in the 16th century.

Physical measurements showing that the Earth does spin have now been made, however. If the size of the Earth is accurately known – it is a sphere 7,926 miles (12,756 km) in diameter – it is simple to calculate the circumference of the planet at the equator. This works out at 24,900 miles (40,074 km). So if the Earth rotates once every 24 hours,

Star trails (below) clearly show the rotation of the Earth. A fixed camera pointed at the sky, with its shutter open for some hours, records the stars as curved trails of light. The center of the circle that the stars trace out is the celestial pole. As the stars are fixed in the sky, it is the spin of the Earth that gives the stars their trails on a long-exposure photo. In this picture of southern hemisphere stars, the camera was pointing southwest over the dome of the Anglo-Australian Telescope.

The curved shapes of weather systems can be seen clearly in a satellite photograph (right) as the spiral of lines of clouds around a low-pressure area.

In low-pressure systems, warm air is rising, so air spirals in to replace the air that rises. In high pressure systems, cool air is descending and displaces air from the center, which then spirals out.

All the Earth's weather is caused by the fact that the Sun gives more heat to the equator than to regions farther north or south. The winds and weather systems act to redistribute excess heat away from the equator. Typically, the equator has many low-pressure systems because the air there is rising, having been heated by the Sun.

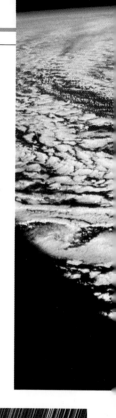

The weather reveals the Earth's rotation because the spin on the planet makes the world's wind-driven weather systems look like giant spirals (**bottom** and **left**). When wind blows from high to low air-pressure areas, the wind's speed has two components: the speed caused by the difference in pressure and the speed of the rotation of the Earth at the point of origin of the wind.

At the equator, the Earth spins fastest, with the spin speed decreasing toward the poles. So when wind blows either north or south, it has the spin speed it picked up at the point of its origin. But the speed of the spin of the Earth below it changes gradually so the path of the wind is curved (**below**). This is called the Coriolis effect. Hence, air spirals counterclockwise around weather systems in the northern hemisphere and clockwise in the southern.

North Pole

Unaffected path

Coriolis force deflection

Rotation of Earth

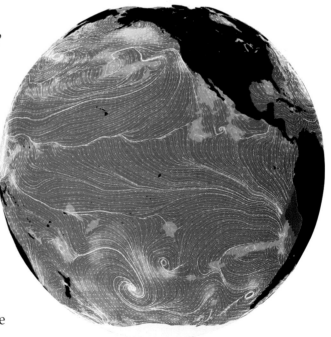

the equator is moving at just over 1,000 mph (1,600 km/h). Why then, if our world is spinning like a turbocharged merry-go-round, do people at the equator not get thrown off? People are held on, of course, by the force of gravity. However, at the equator, people do experience a barely measurable centrifugal force due to this speed, and this makes an average person only about 1 ounce (28 g) lighter than at the poles.

If the Earth moves so rapidly, why can we not feel it moving? From jet travel, we know that just flying steadily at, say, 500 mph (800 km/h) gives us no sensation of movement at all. It is only when the plane banks, accelerates, or slows down that we feel anything. So only a change of speed or direction is detectable, and the Earth spins so steadily that no changes in speed or direction can be perceived.

The rotation of the Earth can actually be witnessed with the help of a long pendulum. A pendulum always swings along a straight line unless there is some force acting on it. With a Foucault pendulum – specially designed so that it is neither subject to any stray external forces, such as air currents, nor affected by any built-in faults, such as a string that untwists – the pendulum's path changes over a few hours. This is because the pendulum, once it has been set swinging, remains moving in a straight line but the Earth has actually rotated beneath it.

Our Earth in orbit

The shadows cast by the Sun can reveal a great deal about the Earth's path around its star.

It is common knowledge that the Sun rises in the east and sets in the west, but few realize that it only does this precisely – rises due east and sets due west – on just two days in the year: the two equinoxes. The rest of the time it rises and sets either to the north or to the south of due east and west.

If you were to mark the spot where it rose or set at midsummer, and do the same at midwinter, it would have moved along a large part of the horizon. The Sun's movements are explained by the fact that the Earth's axis of rotation is tilted from the perpendicular to its orbit, and that the Earth goes around the Sun. The Earth's axial tilt to this perpendicular is 23½ degrees.

In winter when it is observed from the mid-latitudes, the Sun rises late, never gets very high in the sky, and sets early. In summer, of course, the opposite happens. And while the northern hemisphere is enjoying summer, it is winter in the southern hemisphere, and vice versa. It is this axial tilt of 23½ degrees that gives us our seasons.

If the stars were visible during the day, the Sun would appear to move eastward against the background of stars through the year. This happens because the Earth moves around the Sun, so we would see it against a different background month by month. The path that the Sun appears to follow against the sky is known as the ecliptic, which covers all 12 signs of the zodiac. After a year moving around the ecliptic, the Sun is back where it started.

THE FOUR SEASONS

The Earth's axis of rotation points in the same direction throughout the year, but at an angle of 23½ degrees from the perpendicular to its orbit. This gives rise to the seasons – in June the Sun is highest in the sky in the northern hemisphere, while in December it is highest in the southern. On one day in March and September, all parts of the world have days and nights of exactly 12 hours. At all other times, the lengths of day and night are different depending on where you are. In temperate regions, the winter nights are long and the days are short, and vice versa in the summer.

These changes can be witnessed on a sundial and the shadow cast by its gnomon: a high Sun will cast a short shadow. The length of the shadow fluctuates as the Sun's height in the sky changes throughout the year. Even the most accurately constructed sundial will be up to 16 minutes fast or slow at different times of the year. This is because the Earth's orbit is slightly elliptical in shape, not a perfect circle.

For a sundial with an upright gnomon (below), a summer Sun (a) would throw a shadow (a) at a different angle to the shadow (b) of a winter Sun (b) and thus not tell the correct time. This is rectified by using a sloping gnomon. A sundial's gnomon should point north-south and slope at the same angle as the latitude in degrees of the place on the Earth's surface where it is set up.

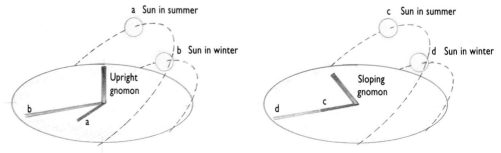

A vivid way of showing how the Sun's elevation in the sky changes throughout the year is to take a photograph using a fixed camera every few days at exactly the same time as shown by a clock, and superimpose the images. In this northern hemisphere example (**left**), a photograph was taken at 8.30 a.m. throughout the year. In winter the Sun is low in the sky, and as the year progresses it moves higher but farther eastward. The figure eight shape, known as an analemma, is an effect caused by the Earth's elliptical orbit which makes the Sun fast or slow compared with the clock. The lower loop is more elongated than the upper because the Earth's orbital speed is greater when it is closer to the Sun during the northern winter. The upper and lower bright lines are trails left by the rising Sun at the solstices. The middle line is the rising Sun's trail at the equinox.

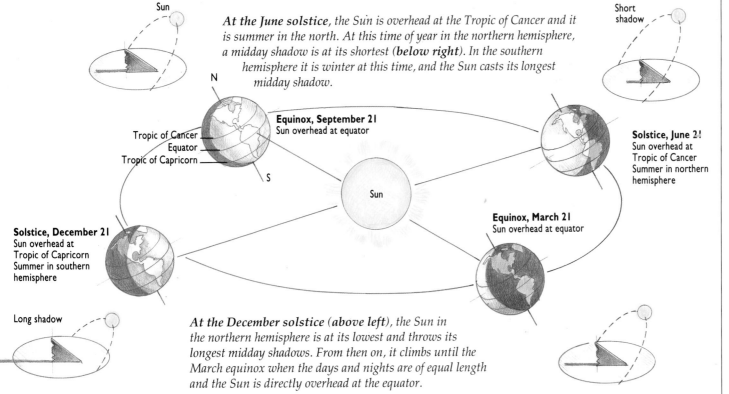

At the June solstice, the Sun is overhead at the Tropic of Cancer and it is summer in the north. At this time of year in the northern hemisphere, a midday shadow is at its shortest (**below right**). In the southern hemisphere it is winter at this time, and the Sun casts its longest midday shadow.

Sun

Short shadow

N

Tropic of Cancer
Equator
Tropic of Capricorn

S

Equinox, September 21
Sun overhead at equator

Solstice, June 21
Sun overhead at Tropic of Cancer
Summer in northern hemisphere

Sun

Solstice, December 21
Sun overhead at Tropic of Capricorn
Summer in southern hemisphere

Equinox, March 21
Sun overhead at equator

Long shadow

At the December solstice (above left), the Sun in the northern hemisphere is at its lowest and throws its longest midday shadows. From then on, it climbs until the March equinox when the days and nights are of equal length and the Sun is directly overhead at the equator.

Celestial landmarks

Individual stars are easy to find once you have learned how to recognize a few key constellations and use them as signposts.

On a clear night out in the country, the sky is so full of stars that it can seem an impossible task to learn which constellation is which. But finding your way around the sky is not as hard as it seems. It is best to learn a few landmarks and go from there – as you would if you were finding your way around a new city. In the sky there are a few constellations that are easy to recognize and can act as celestial signposts to the rest of them.

It is best in fact to start your celestial "journey" on a rather poor night – the sort you might get when it is a little misty, or in town where there is a lot of "light pollution" from street lights. In these conditions, only the brighter stars are visible, and you can pick out the patterns, or constellations, more easily. The eye is good at seeing patterns, and after a while you will learn the configurations of some of the major, more prominent constellations.

From there, it is only a short step to discovering the more complex patterns found throughout a clear night sky.

*The most useful landmark in northern skies is Ursa Major, the Great Bear (**below**). Its seven brightest stars – from Alkaid to Dubhe – make up the Big Dipper (or Plow), a shape that is easy to recognize. As a pointer, it is hard to beat. For instance, follow the two right-hand stars upward and you arrive at Polaris, the North Star, in Ursa Minor, the Little Bear. This star marks the north celestial pole.*

If you extend the curve of the handle downward from Alkaid, you find the constellation of Boötes, containing the bright star Arcturus. Look in the opposite direction and you pass through the Lynx before arriving at Castor and Pollux in Gemini. To the south lies Leo, containing Regulus and Algieba.

USING A PLANISPHERE

A planisphere is a circular sky map with a rotating mask that shows you which stars are above the horizon at any time. Each planisphere is designed for a specific latitude on the Earth, plus or minus a few degrees. Even experienced observers find one useful to help them figure out when to look for a particular object.

Around the edge are a series of dates, and on the rotating mask are times. After simply matching the date with the time, the clear area in the mask then shows the entire sky visible at that moment. The edge of the mask is the horizon. To use a planisphere, hold it above and in front of you so you look up to see the mask with the center of the disk toward the pole. The stars on the map are then in the correct orientation for the chosen time and date.

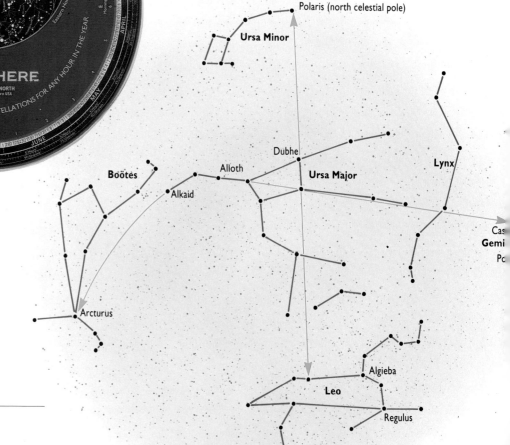

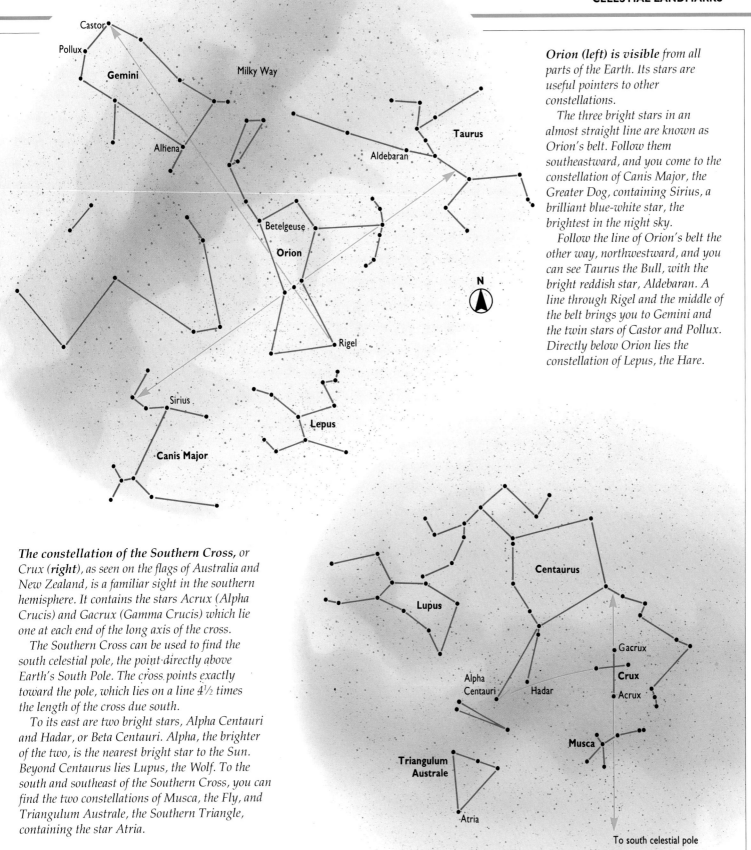

Orion (left) is visible from all parts of the Earth. Its stars are useful pointers to other constellations.

The three bright stars in an almost straight line are known as Orion's belt. Follow them southeastward, and you come to the constellation of Canis Major, the Greater Dog, containing Sirius, a brilliant blue-white star, the brightest in the night sky.

Follow the line of Orion's belt the other way, northwestward, and you can see Taurus the Bull, with the bright reddish star, Aldebaran. A line through Rigel and the middle of the belt brings you to Gemini and the twin stars of Castor and Pollux. Directly below Orion lies the constellation of Lepus, the Hare.

The constellation of the Southern Cross, or Crux (**right**), as seen on the flags of Australia and New Zealand, is a familiar sight in the southern hemisphere. It contains the stars Acrux (Alpha Crucis) and Gacrux (Gamma Crucis) which lie one at each end of the long axis of the cross.

The Southern Cross can be used to find the south celestial pole, the point directly above Earth's South Pole. The cross points exactly toward the pole, which lies on a line 4½ times the length of the cross due south.

To its east are two bright stars, Alpha Centauri and Hadar, or Beta Centauri. Alpha, the brighter of the two, is the nearest bright star to the Sun. Beyond Centaurus lies Lupus, the Wolf. To the south and southeast of the Southern Cross, you can find the two constellations of Musca, the Fly, and Triangulum Australe, the Southern Triangle, containing the star Atria.

17

Viewing the sky

For more than three centuries, people have used telescopes to look at the sky.

A telescope collects more light than normally enters your eye, so that you can see faint objects and receive a magnified view giving more detail. The task of collecting the light is carried out either by a lens or by a mirror. These bend light rays from a distant object and focus them to form an image. To magnify this, another lens, the eyepiece, is used. Almost all astronomical telescopes give an upside-down image. To correct this would need extra mirrors or lenses. But as this would involve the loss of some light, it is avoided.

A refracting telescope (below and right) uses its object lens to gather and focus light at the focal point. There, it is magnified by the eyepiece. The photo shows a 3¼-inch (80-mm) refractor, in other words its lens is 3¼-inches (80-mm) in diameter. It is on an equatorial mounting. This type of telescope is a popular choice for the amateur astronomer, providing good images of the planets.

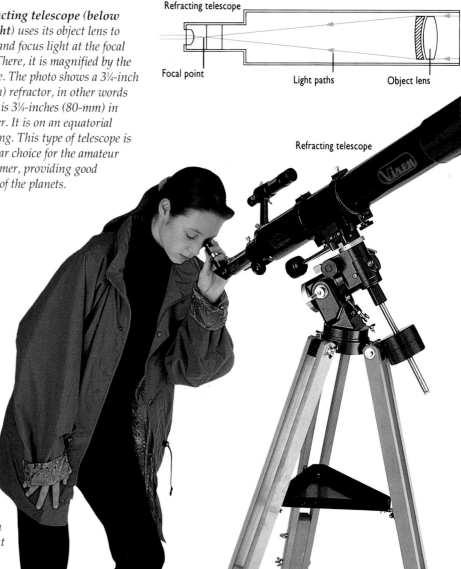

Refracting telescope

Focal point · Light paths · Object lens

Refracting telescope

Schmidt–Cassegrain telescope

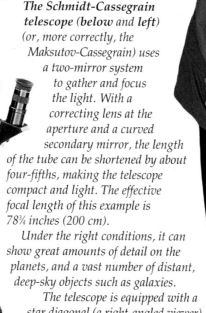

The Schmidt-Cassegrain telescope (below and left) (or, more correctly, the Maksutov-Cassegrain) uses a two-mirror system to gather and focus the light. With a correcting lens at the aperture and a curved secondary mirror, the length of the tube can be shortened by about four-fifths, making the telescope compact and light. The effective focal length of this example is 78¾ inches (200 cm).

Under the right conditions, it can show great amounts of detail on the planets, and a vast number of distant, deep-sky objects such as galaxies.

The telescope is equipped with a star diagonal (a right-angled viewer). This allows the astronomer to use the telescope with comparative ease at otherwise awkward angles.

Schmidt–Cassegrain telescope

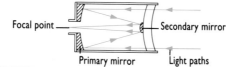

Focal point · Secondary mirror

Primary mirror · Light paths

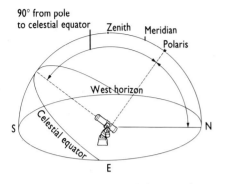

An equatorial mount (above and left) allows objects to be followed easily as they move across the sky.

To set up an equatorial mount, you must set the polar axis of the mount at the latitude of the observing site and aim it at the celestial pole, which in the northern hemisphere is near Polaris. When this is done, and the telescope is aimed at an object in the sky, say, on the celestial equator 90° from the celestial pole, it only has to be rotated around its polar axis to follow the object as the sky rotates from east to west.

A Dobsonian mount (below) is a simple altazimuth mount used to support a Newtonian reflecting telescope. An altazimuth mount moves up and down and from side to side; to track an object it has to make two separate movements.

Newtonian reflector on Dobsonian altazimuth mounting

BINOCULARS

Many amateurs use binoculars as much as their telescopes. They can show a number of nebulae and star clusters and can even be used for observing objects such as variable stars and for comet spotting.

A pair of binoculars is, in effect, two refracting telescopes side by side, allowing both eyes to be used. Binoculars are described by two figures, for instance 7 x 50, meaning that the magnification is 7 and the main lenses are 50 mm (2 inches) across. Mounts are available to allow you a steady view of the sky.

In porro prism binoculars (above and left), the most popular and convenient type, prisms bend the path of the light. This both keeps the tube length down, and gives an upright image.

Porro prism binoculars
Eyepiece lens
Porro prism
Light path
Object lens

The magnification given by a telescope can be worked out by dividing the focal length of the mirror or lens – the distance from the mirror or lens to its focal point – by the focal length of the eyepiece. The more an image is magnified, the fainter the image, since the same amount of light is spread over a larger area.

The amount of detail that any mirror or lens will show is limited. To get a brighter, more magnified image, a larger aperture is needed. A larger lens or mirror will do this. Unsteadiness will blur any image, so a firm stand and a stable mounting are essential.

Newtonian reflecting telescope

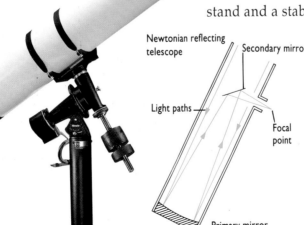

Newtonian reflecting telescope
Secondary mirror
Light paths
Focal point
Primary mirror

Newtonian reflecting telescopes (left) use a curved mirror to collect and reflect light. The light is focused onto a secondary mirror and reflected to the eyepiece on the side of the telescope. The eyepiece is situated at the focal point to magnify the image produced.

The design of a reflector means that air currents can form within the tube, creating slightly imperfect images, but a latticed tube can alleviate this in larger instruments.

The home observatory

With a fully equipped observatory or just a pair of binoculars, amateurs can make useful records.

The simplest form of astronomy, stargazing, can be practiced using just the naked eye. But people who want to see a little more of the wonders of the universe use either a telescope or binoculars. A telescope should be used outdoors to avoid the heat from the house that creates air turbulence and distorts the image. Telescopes should also be set up away from nearby lights to avoid light pollution – artificial lights illuminating the sky and making faint objects hard to see. A typical nighttime session usually includes a look at any planets that are in the sky followed by searches for distant objects.

Many people take up astronomy just for fun, but because professional astronomers cannot maintain a constant watch on the whole sky, amateurs can play a part by making useful observations. With the help of a reflecting telescope of 6-inch (15-cm) or more aperture, detailed drawings can be made of the planets. Amateurs often spot such events as new outbreaks of activity in either Jupiter's or Saturn's clouds, or new dust storms on Mars. Others make estimates of the brightness of variable stars, even using an ordinary pair of binoculars. Less equipment is needed for meteor observing. Simply watching the sky and making notes of the brightness and source of every shooting star is a help.

JUPITER SECTION
B.A.A.

JUPITER SECTION
B.A.A.

DATE October 7 1988 U.T. 22:45
LONGITUDE OF C.M. SYSTEM 1 327.6°
LONGITUDE OF C.M. SYSTEM 11 286.4°
INSTRUMENT 157mm spec
SEEING 2
OBSERVER Matthew J. Boulton
(FOR NOTES SEE BACK)
STB Ovals "BC" and "DE" divided by
STB remnant. Ganymede reappears
following occultation

JUPITER SECTION
B.A.A.

DATE 24 November 1988 U.T. 20:54
LONGITUDE OF C.M. SYSTEM 1 285.4°
LONGITUDE OF C.M. SYSTEM 11 238.6°
INSTRUMENT 215mm
SEEING A III
OBSERVER D. Fisher
(FOR NOTES SEE BACK)
Shows activity in the equatorial
zone adjacent the NEB

In addition to making informative sketches of the object being observed (**left**), it is vital to keep comprehensive notes. Their content will depend on what is being observed, but for a planet such as Jupiter, these notes might include the following: the date; the time in universal time (U.T., the same as Greenwich mean time); a reference to what part of the planet is being observed (C.M. refers to the central meridian of Jupiter); the size of the aperture of the instrument being used; an assessment of the seeing (the steadiness of the atmosphere) at the time of the observation; the observer's name; and any notes or points of special interest about the object.

CAPTURING THE IMAGE

Some advanced amateur astronomers use charge-coupled devices (CCDs). These are microchips, similar to those found in video cameras, which contain thousands of tiny photo-sensitive pixels. These generate a small charge when exposed to even low levels of light. The charges are converted into digital signals, so an image can be displayed on a computer screen.

CCDs allow the amateur to capture images of faint objects, once only possible with long-exposure photographs and large professional telescopes.

Only long-exposure photos can reveal fine detail on faint objects. When there is light pollution, exposure time must be kept short or stray light fogs the film. So a photo taken with even a medium-sized telescope, such as a 10-inch (25-cm) reflector, will not show much detail (**above**).

A view of the same part of the sky using a CCD reveals far more detail (**above**). In this instance, the galaxy M81 appears, whereas before there had only been smudges of light. The short exposure time needed with a CCD reduces the effects of light pollution.

The image (below), taken in March 1986, *shows the galaxy Centaurus A (NGC 5128), prior to the eruption of supernova 1986G. Supernovae are labeled in alphabetical order – 1986G was the 7th supernova (G being the 7th letter of the alphabet) to be discovered in 1986.*

Amateur astronomers *make an invaluable contribution to the discovery of objects such as supernovae and comets. One particularly successful amateur, the Australian Robert Evans (above), had, by the year end of 1993, discovered 26 supernovae in distant galaxies, of which 5 were shared discoveries with other amateurs.*

Astronomers do not need sophisticated equipment. English amateur George Alcock has discovered five comets and six novae merely using binoculars.

As it erupted, *the supernova 1986G (left, marked by the arrow) was discovered by the amateur astronomer Robert Evans. Once notified of its existence, observatories then turned their attention to the phenomenon.*

With a medium-sized telescope, *detailed features on the Moon become visible. The best time to view a part of the Moon is when the Sun is shining across it at an angle. The shadows thrown across the surface emphasize the features, making them easier to recognize. Fracastorius (right) is a medium-sized crater in the southeast corner of the near side of the Moon. This amateur drawing shows how much detail can be seen. A well-labeled drawing is essential to allow others to benefit from observations.*

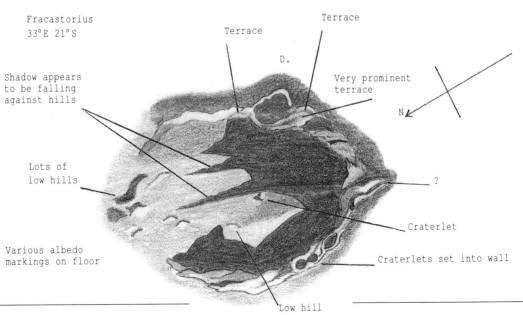

Fracastorius
33° E 21° S

Terrace

Terrace

D.

Very prominent terrace

N

Shadow appears to be falling against hills

Lots of low hills

?

Craterlet

Various albedo markings on floor

Craterlets set into wall

Low hill

Mapping the sky

There have been astronomers ever since humans first looked up at the sky.

The word "astronomer" simply means "star-namer." Some of the names of stars today may have been used by the peoples of the Near East as long ago as 3000 B.C. In the earliest days, people must have looked at the skies and made up stories which explained the patterns they saw there.

Few of the constellations, or patterns of stars, look anything like the objects that their names suggest, but they have remained with us because dividing up the sky makes it easier to recognize. A great number of the brighter stars have names – many given to them by Arab astronomers. Often these names refer to the parts of the figures that the stars represent. Rigel in Orion, for example, comes from the Arabic *Rijl al Jawzah*, meaning "Leg of the Giant."

A more scientific method of listing the stars emerged in 1603, when the German astronomer Johann Bayer assigned the letters of the Greek alphabet to the stars in each constellation, in order of brightness, starting with alpha and ending with omega. In this scheme, Bellatrix is known as Gamma Orionis – Orionis is the Latin possessive form of Orion and means "of the Hunter." In other words, it is the third brightest star in the constellation of Orion.

This system lasted until 1725, when a catalog produced by the English Astronomer Royal, John Flamsteed, listed the stars of each constellation in the order of their right ascension. Today, a variety of star catalogs exist that are used to account for the increasingly faint stars that are being found. Among other catalogs is one called the Struve catalog, which lists double stars.

This section of a star atlas (above) compiled in the 17th century shows the traditional constellation figures superimposed on the stars. The star patterns are mirror images of what we see in the sky because the atlas is made to match a celestial globe. This is similar to an Earth globe, but it shows the sky as if we were on the outside looking in.

Many new constellations were named in the 18th century, when gaps between the traditional constellations were filled in and the previously uncharted southern skies were mapped. Some were named for what were then the wonders of science, such as the air pump, Antlia.

The celestial sphere (right) is an imaginary globe around the Earth, upon which the night sky is projected. Over small areas of the sky, this gives a two-dimensional impression of the heavens, even though each star may be thousands of light-years farther away or closer to Earth than others in the constellation.

The sphere is divided into gradations of right ascension and declination, in much the same way as the Earth is divided up into latitude and longitude, providing coordinates for objects in the sky. Declination sets an object's height in degrees above or below the celestial equator, and right ascension determines how far around the equator it is in units of hours and minutes.

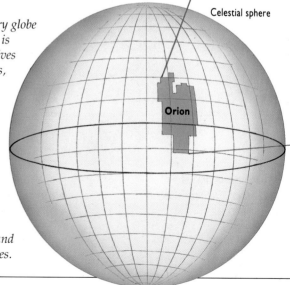

Celestial sphere

Right ascension

6 hours

5 hours

λ
φ
α
Betelgeuse
Bellatrix
γ
+5°
ψ
Orion
0
δ
Mintaka
ε
Alnitak **Alnilam**
ζ
τ
−5°
β
Rigel
κ
−10°
Saiph

Declination

- ● 1st magnitude
- ✪ 2nd magnitude
- ★ 3rd magnitude
- • 4th magnitude
- · 5th magnitude

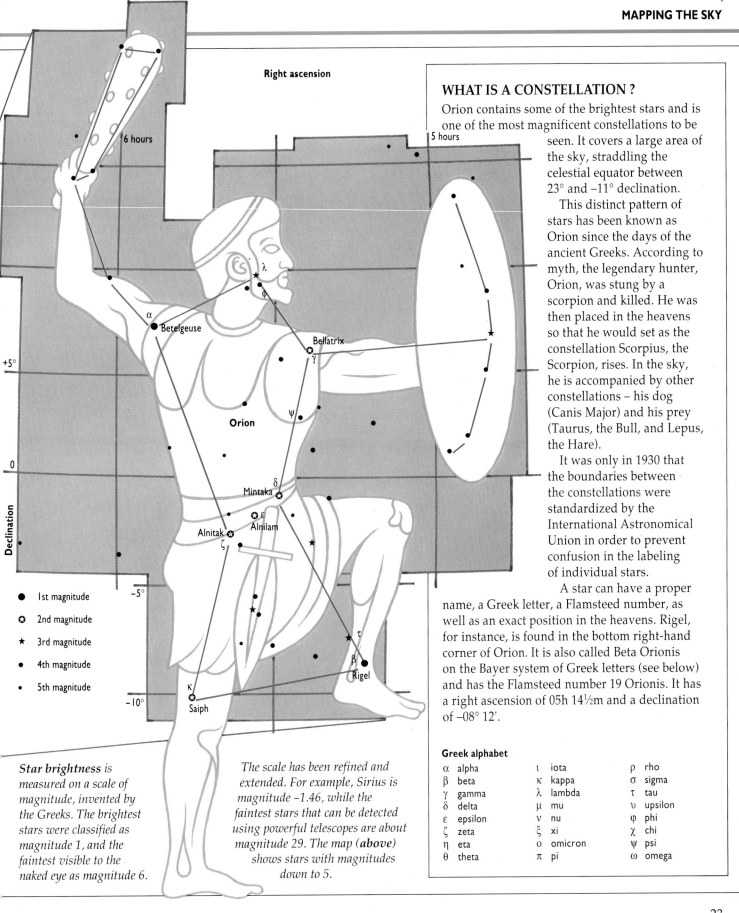

WHAT IS A CONSTELLATION ?

Orion contains some of the brightest stars and is one of the most magnificent constellations to be seen. It covers a large area of the sky, straddling the celestial equator between 23° and −11° declination.

This distinct pattern of stars has been known as Orion since the days of the ancient Greeks. According to myth, the legendary hunter, Orion, was stung by a scorpion and killed. He was then placed in the heavens so that he would set as the constellation Scorpius, the Scorpion, rises. In the sky, he is accompanied by other constellations – his dog (Canis Major) and his prey (Taurus, the Bull, and Lepus, the Hare).

It was only in 1930 that the boundaries between the constellations were standardized by the International Astronomical Union in order to prevent confusion in the labeling of individual stars.

A star can have a proper name, a Greek letter, a Flamsteed number, as well as an exact position in the heavens. Rigel, for instance, is found in the bottom right-hand corner of Orion. It is also called Beta Orionis on the Bayer system of Greek letters (see below) and has the Flamsteed number 19 Orionis. It has a right ascension of 05h 14½m and a declination of −08° 12'.

Star brightness is measured on a scale of magnitude, invented by the Greeks. The brightest stars were classified as magnitude 1, and the faintest visible to the naked eye as magnitude 6.

*The scale has been refined and extended. For example, Sirius is magnitude −1.46, while the faintest stars that can be detected using powerful telescopes are about magnitude 29. The map (**above**) shows stars with magnitudes down to 5.*

Greek alphabet

α	alpha	ι	iota	ρ	rho
β	beta	κ	kappa	σ	sigma
γ	gamma	λ	lambda	τ	tau
δ	delta	μ	mu	υ	upsilon
ε	epsilon	ν	nu	φ	phi
ζ	zeta	ξ	xi	χ	chi
η	eta	ο	omicron	ψ	psi
θ	theta	π	pi	ω	omega

Skymap 1

To show the positions of the constellations and stars, the sky is divided into maps.

The maps on the following pages (pp. 26–29) show stars of the fifth magnitude or brighter. The names of the constellations appear in capital letters, and their boundaries are shown as straight, broken (dotted) lines. The names of the most important stars in the constellations are in capitals and small letters, and red lines join up a constellation's brightest stars. The cloudlike regions show where our galaxy appears in the sky, but the Galaxy can only be seen when the sky is clear and there is little light pollution. The only moving object whose path is shown here is the Sun, which over the course of a year appears to move eastward around the sky along a path known as the ecliptic.

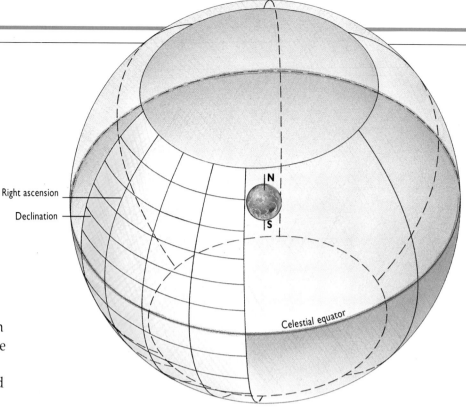

Right ascension

Declination

Celestial equator

There are 88 constellations (right) – each has a Latin name, by which it is known to astronomers around the world, as well as a common name.

The coordinates used in the table are for a position approximately in the center of each constellation. So to find a constellation, first look up its map number, then use the R.A. (right ascension) and Decl. (declination) coordinates to locate the middle of the constellation.

When stars in a constellation are referred to by Greek letters or other classifications, such as their Flamsteed numbers, the convention is that the name of the constellation alters slightly. It is used in the Latin possessive form, meaning "belonging to," which changes the ending of the name of the constellation. Thus the star α in Orion becomes α Orionis.

LIST OF THE CONSTELLATIONS

Name English name	R.A. h m	Decl. °	Map/s
Andromeda Andromeda	01 00	+40	1/2/8
Antlia The Air Pump	10 20	−34	6/7
Apus The Bird of Paradise	16 00	−76	5
Aquarius The Water Carrier	22 30	−10	2/3
Aquila The Eagle	19 40	+05	3
Ara The Altar	17 20	−52	4/5
Aries The Ram	02 40	+20	8
Auriga The Charioteer	05 30	+40	1/7/8
Boötes The Herdsman	14 40	+30	1/4/6
Caelum The Engraving Tool	04 50	−38	8
Camelopardus The Giraffe	05 00	+70	1
Cancer The Crab	08 40	+20	7
Canes Venatici The Hunting Dogs	13 00	+40	1/6
Canis Major The Greater Dog	06 50	−22	7
Canis Minor The Lesser Dog	07 40	+05	7
Capricornus The Goat	21 00	−18	3
Carina The Keel	09 00	−60	5
Cassiopeia Cassiopeia	01 00	+60	1
Centaurus The Centaur	13 00	−48	4/5/6
Cepheus Cepheus	22 30	+70	1
Cetus The Whale	01 40	−05	2/8
Chamaeleon The Chameleon	10 30	−80	5

Name English name	R.A. h m	Decl. °	Map/s
Circinus The Compasses	15 00	−62	5
Columba The Dove	05 40	−35	7/8
Coma Berenices Berenice's Hair	12 50	+25	6
Corona Australis The Southern Crown	18 40	−42	3
Corona Borealis The Northern Crown	16 00	+30	4
Corvus The Crow	12 30	−20	6
Crater The Cup	11 20	−15	6
Crux The Southern Cross	12 30	−60	5
Cygnus The Swan	20 40	+40	1/3
Delphinus The Dolphin	20 40	+12	3
Dorado The Goldfish (Swordfish)	05 00	−60	5
Draco The Dragon	16 20	+62	1
Equuleus The Little Horse	21 10	+08	3
Eridanus The River/River Eridanus	04 00	−10	5/8
Fornax The Furnace	02 40	−32	2/8
Gemini The Twins	07 00	+25	7
Grus The Crane	22 30	−45	2/5
Hercules Hercules	17 30	+30	3/4
Horologium The (Pendulum) Clock	03 20	−52	5
Hydra The Water Snake (Sea Serpent)	10 00	−20	4
Hydrus The Lesser Water Snake	02 00	−72	5/6/7
Indus The Indian	21 30	−56	3/5

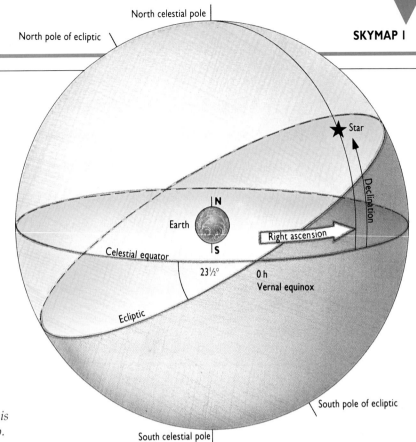

Like the latitude and longitude system on terrestrial maps, there is a coordinate system for the sky (right). The coordinates are projected onto the celestial sphere (left), an imaginary sphere around Earth on which the stars appear. The equivalent of latitude, known as declination, is measured in degrees and is linked to the Earth's own latitudes. Stars of 0° declination are above Earth's equator, while a star of 90° is above the North Pole. Declinations in the southern half of the celestial sphere have negative values.

The equivalent of longitude is right ascension (R.A.) and is measured in hours, minutes, and seconds. The R.A. scale has 24 hours and runs from west to east, starting from the point where the ecliptic crosses the equator at the vernal equinox, in other words, where the Sun crosses the equator in the month of March. This point is determined by where the Earth's axis is pointing. Since the direction the axis points in slowly moves in a circle once every 28,500 years, a process called precession, this means that star charts need to be revised every 50 years or so.

Name English name	R.A. h m	Decl. °	Map/s
Lacerta The Lizard	22 30	+45	1/2
Leo The Lion	10 30	+20	6
Leo Minor The Lesser Lion	10 20	-32	6/7
Lepus The Hare	05 30	-18	7/8
Libra The Scales	14 10	-15	4
Lupus The Wolf	15 20	-42	4/5
Lynx The Lynx	08 00	+46	1/7
Lyra The Lyre	18 50	+36	3
Mensa The Table (Mountain)	05 30	-75	5
Microscopium The Microscope	21 00	-36	3
Monoceros The Unicorn	07 00	00	7
Musca The Fly	12 30	-70	5
Norma The Level	16 00	-52	4/5
Octans The Octant	21 00	-85	5
Ophiuchus The Serpent Bearer	17 20	-05	3/4
Orion Orion	05 30	00	7/8
Pavo The Peacock	19 20	-65	5
Pegasus Pegasus	22 40	+20	2/3
Perseus Perseus	03 40	+45	1/8
Phoenix The Phoenix	00 50	-50	2/5
Pictor The (Painter's) Easel	05 30	-54	5/8
Pisces The Fishes	00 30	+14	2

Name English name	R.A. h m	Decl. °	Map/s
Piscis Austrinus The Southern Fish	22 30	-31	2/3
Puppis The Stern	07 40	-34	7
Pyxis The Mariner's Compass	09 00	-28	7
Reticulum The Net	03 50	-62	5
Sagitta The Arrow	19 40	+19	3
Sagittarius The Archer	19 00	-28	3/4
Scorpius The Scorpion	17 00	-32	4
Sculptor The Sculptor('s Workshop)	00 30	-32	2
Scutum The Shield	18 40	-10	3
Serpens The Serpent	15 40	+10	3/4
Sextans The Sextant	10 20	00	6/7
Taurus The Bull	04 30	+15	8
Telescopium The Telescope	19 20	-52	3/5
Triangulum The Triangle	02 10	+32	2
Triangulum Australe The Southern Triangle	16 00	-65	5
Tucana The Toucan	00 00	-65	5
Ursa Major The Great Bear (Big Dipper)	11 00	+55	1/6/7
Ursa Minor The Little Bear (Little Dipper)	15 00	+75	1
Vela The Sails	09 40	-48	5/6/7
Virgo The Virgin	13 20	00	4/6
Volans The Flying Fish	08 00	-70	5
Vulpecula The Fox	20 20	+25	3

α alpha
β beta
γ gamma
δ delta
ε epsilon
ζ zeta
η eta
θ theta
ι iota
κ kappa
λ lambda
μ mu
ν nu
ξ xi
ο omicron
π pi
ρ rho
σ sigma
τ tau
υ upsilon
φ phi
χ chi
ψ psi
ω omega

Skymap 2

The night sky in the northern hemisphere contains the familiar North Star, Polaris, and in Cygnus our galaxy shows rich star fields.

The celestial north pole has a bright star within about one degree of declination of the pole itself. Called, appropriately, Polaris (α Ursae Minoris), the star is a boon to lost travelers in the northern hemisphere. If they can find Polaris, then at least they know where north is. The star is quite easy to find because Merak and Dubhe (β and α Ursae Majoris) clearly point at it. These two stars make up the end of the "ladle" in the constellation that is also known as the Great Bear. The seven prominent stars of Ursa Major are also called the Big Dipper or the Plow.

Another notable constellation is Cassiopeia. The stars ε, δ, γ, α and β Cassiopeiae trace out a clear "W" shape, making it one of the most distinctive constellations of northern skies, especially on winter evenings when it is virtually overhead as viewed from mid-latitudes.

The Square of Pegasus dominates the northern part of this segment of sky. One of the stars, Alpheratz, is shared with Andromeda to the north. In Andromeda is M31, the closest galaxy to our own, visible as a milky pool on very clear nights.

- ● 1st magnitude
- ✪ 2nd magnitude
- ★ 3rd magnitude
- • 4th magnitude
- · 5th magnitude

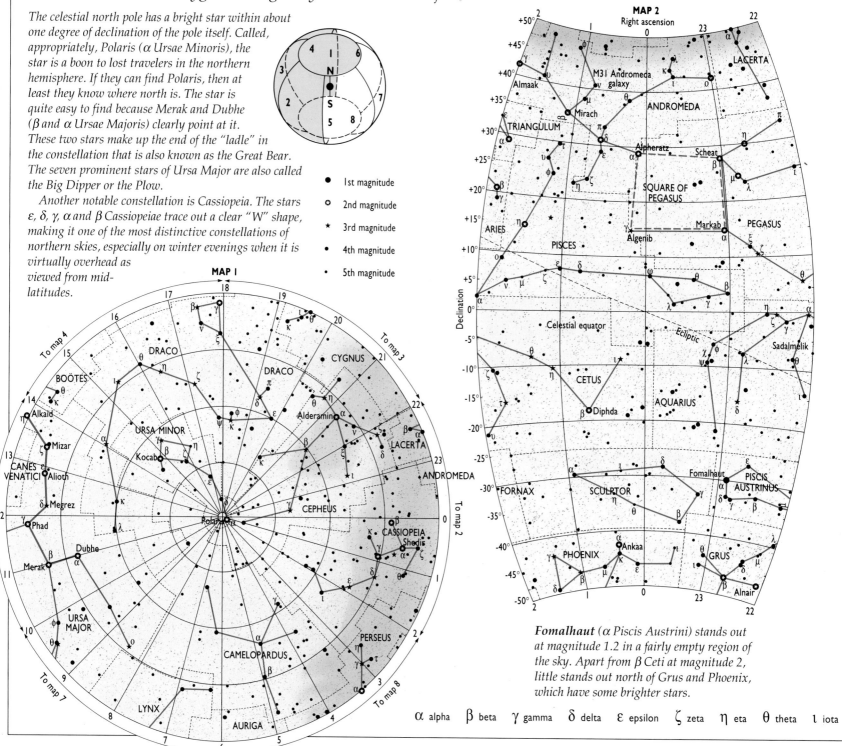

Fomalhaut (α Piscis Austrini) stands out at magnitude 1.2 in a fairly empty region of the sky. Apart from β Ceti at magnitude 2, little stands out north of Grus and Phoenix, which have some brighter stars.

α alpha   β beta   γ gamma   δ delta   ε epsilon   ζ zeta   η eta   θ theta   ι iota

North of the equator *are three fine constellations: Cygnus, Lyra (with Vega, the 5th brightest star in the sky), and part of Aquila (with the 1st magnitude star Altair). The Galaxy in Cygnus looks superb through a pair of binoculars.*

Rasalgethi *(α Herculis) varies from magnitude 3 to 4. And between η and ζ is M13, a globular cluster just visible in dark skies; it is a fine sight through a telescope with an aperture above 8in. (200 mm). In Boötes is the bright, reddish star Arcturus.*

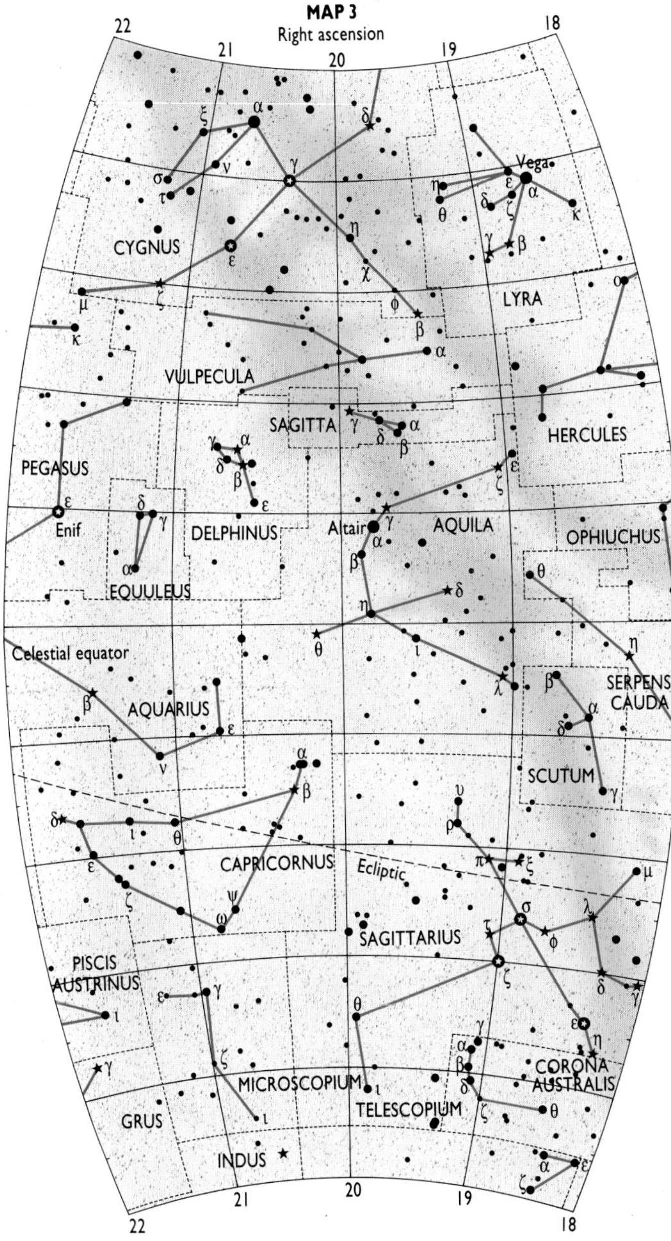

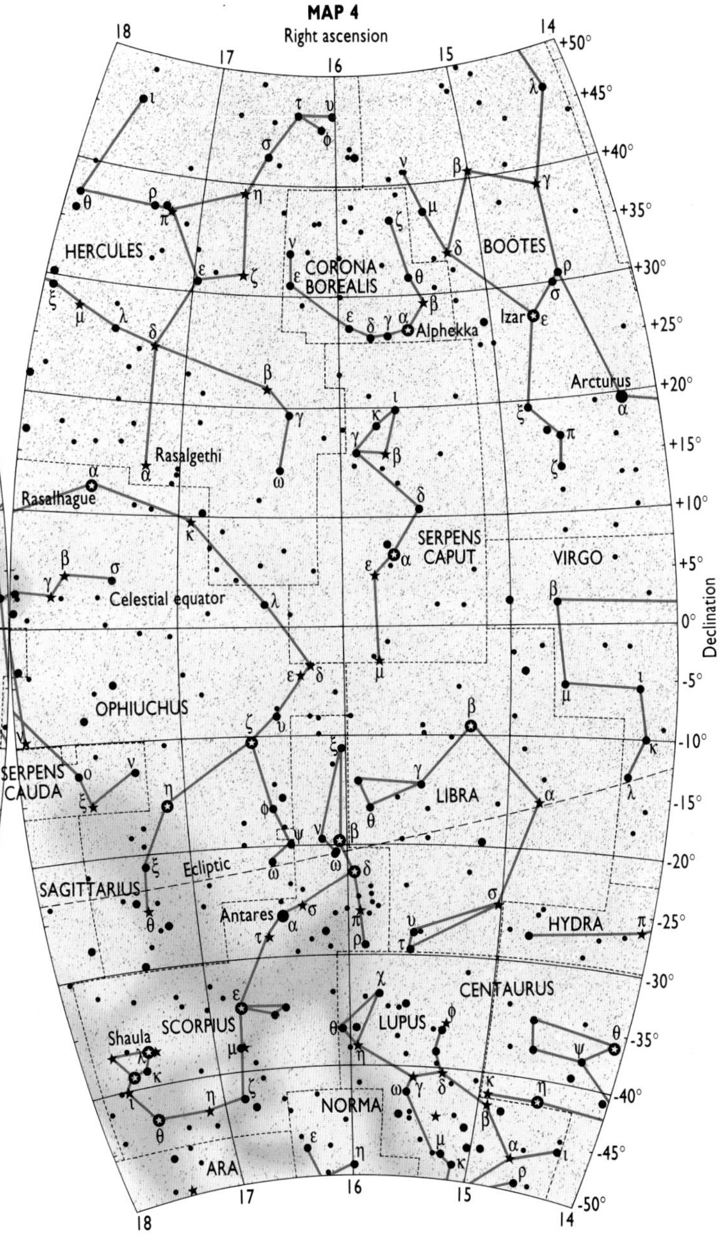

See also

OBSERVATORY EARTH
▶ Celestial landmarks 16/17

▶ Viewing the sky 18/19

▶ Mapping the sky 22/23

▶ Skymaps 24/29

THE PLANETS
▶ Wandering stars 52/53

SUN AND STARS
▶ Star bright 112/113

▶ Double stars 116/117

▶ Variable stars 118/119

NEBULAE AND GALAXIES
▶ Clouds that never rain 132/133

▶ Star clusters 136/137

▶ Islands in space 138/139

▶ Our galaxy 144/145

MAP 3
Right ascension

Constellations shown: CYGNUS, LYRA, VULPECULA, SAGITTA, PEGASUS, DELPHINUS, EQUULEUS, AQUILA, HERCULES, OPHIUCHUS, AQUARIUS, SCUTUM, SERPENS CAUDA, CAPRICORNUS, SAGITTARIUS, PISCIS AUSTRINUS, MICROSCOPIUM, TELESCOPIUM, CORONA AUSTRALIS, GRUS, INDUS. (Vega, Altair, Enif; Ecliptic; Celestial equator)

MAP 4
Right ascension

Constellations shown: HERCULES, CORONA BOREALIS, BOÖTES, SERPENS CAPUT, VIRGO, OPHIUCHUS, LIBRA, SAGITTARIUS, SCORPIUS, LUPUS, CENTAURUS, NORMA, ARA, HYDRA. (Rasalgethi, Rasalhague, Alphekka, Izar, Arcturus, Antares, Shaula; Celestial equator; Ecliptic; Declination)

The south of this section *is dominated by Sagittarius, a zodiacal constellation (one with the ecliptic through it). Our galaxy's center is in Sagittarius, so the area is rich in star fields and a glorious sight in a dark sky.*

Scorpius, *with the bright star Antares (α Scorpii) at magnitude 0.9, dominates the south of this section. Our galaxy is especially rich in this region and is well worth viewing through binoculars.*

κ kappa λ lambda μ mu ν nu ξ xi ο omicron π pi ρ rho σ sigma τ tau υ upsilon φ phi χ chi ψ psi ω omega

Skymap 3

There are many fascinating things to see in the southern night skies, and the constellation Orion stands out on the celestial equator.

The region of the sky around the south celestial pole is packed with interesting objects. One of these is the star Rigil Kentaurus (α Centauri) which, at just over 4.3 light-years, is the second closest star to our Sun and the closest visible with the naked eye. It is also a bright star at magnitude –0.3. The Galaxy is especially rich in the region around Crux and Centaurus, and the star fields look stunning through binoculars.

There are also two nebulae visible in the southern skies – the Small Magellanic Cloud in Tucana and the Large Magellanic Cloud straddling Dorado and Mensa. These "clouds" are two of the galaxies that make up the Local Group of galaxies and are near neighbors of our galaxy at between 160,000 and 180,000 light-years away. They are named after the 16th-century Portuguese navigator Magellan.

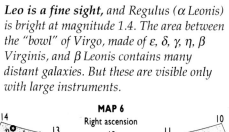

- ● 1st magnitude
- ⊛ 2nd magnitude
- ★ 3rd magnitude
- • 4th magnitude
- · 5th magnitude

Leo is a fine sight, and Regulus (α Leonis) is bright at magnitude 1.4. The area between the "bowl" of Virgo, made of ε, δ, γ, η, β Virginis, and β Leonis contains many distant galaxies. But these are visible only with large instruments.

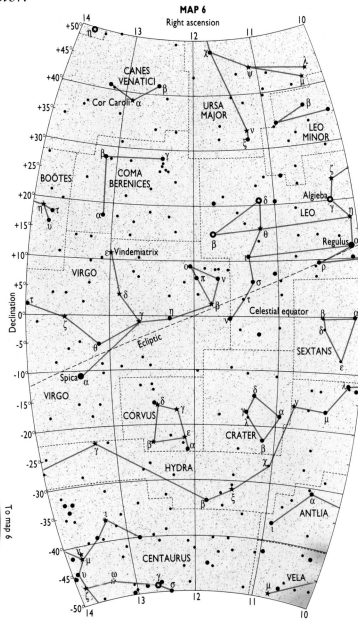

MAP 6

Right ascension

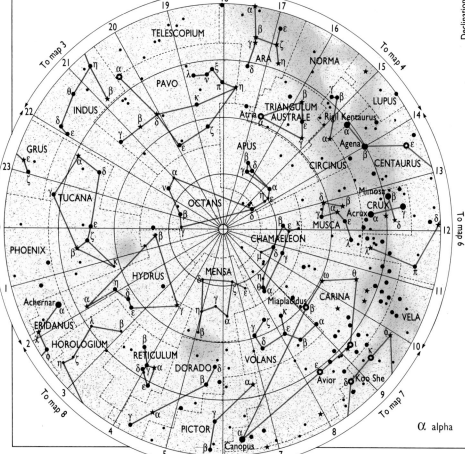

MAP 5

Omega Centauri is not a star, but the best globular cluster in the sky. Spica (α Virginis) is a bright star just below the ecliptic. The main stars of Corvus (the Crow) form a prominent quadrilateral close to Spica.

α alpha β beta γ gamma δ delta ε epsilon ζ zeta η eta θ theta ι iota

***Dominated by Castor and Pollux** at magnitudes 1.6 and 1.2, respectively, Gemini is a fine sight. Castor is a binary; each of its two stars can be distinguished with a small telescope. The stars revolve around their center of mass every 350 years.*

***Taurus has two open clusters,** the Hyades and the Pleiades. Algol (β Persei) "winks" from magnitude 2.3 to 3.5 and back over a five-hour period every 2.86 days. Capella (α Aurigae) is a bright 0.1 magnitude yellowish star and a spectroscopic binary.*

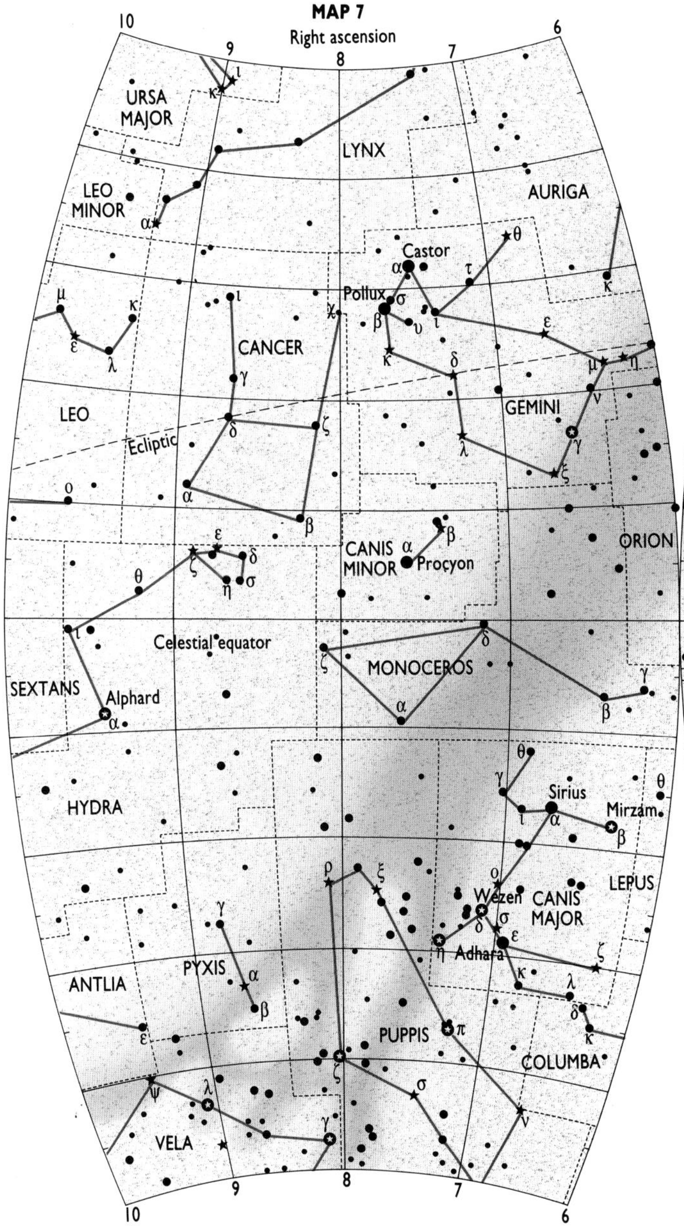
MAP 7
Right ascension
10
9
8
7
6
URSA MAJOR
LYNX
AURIGA
LEO MINOR
Castor
Pollux
CANCER
GEMINI
LEO
Ecliptic
CANIS MINOR
Procyon
ORION
Celestial equator
MONOCEROS
SEXTANS
Alphard
HYDRA
Sirius
Mirzam
LEPUS
Wezen
CANIS MAJOR
Adhara
ANTLIA
PYXIS
PUPPIS
COLUMBA
VELA

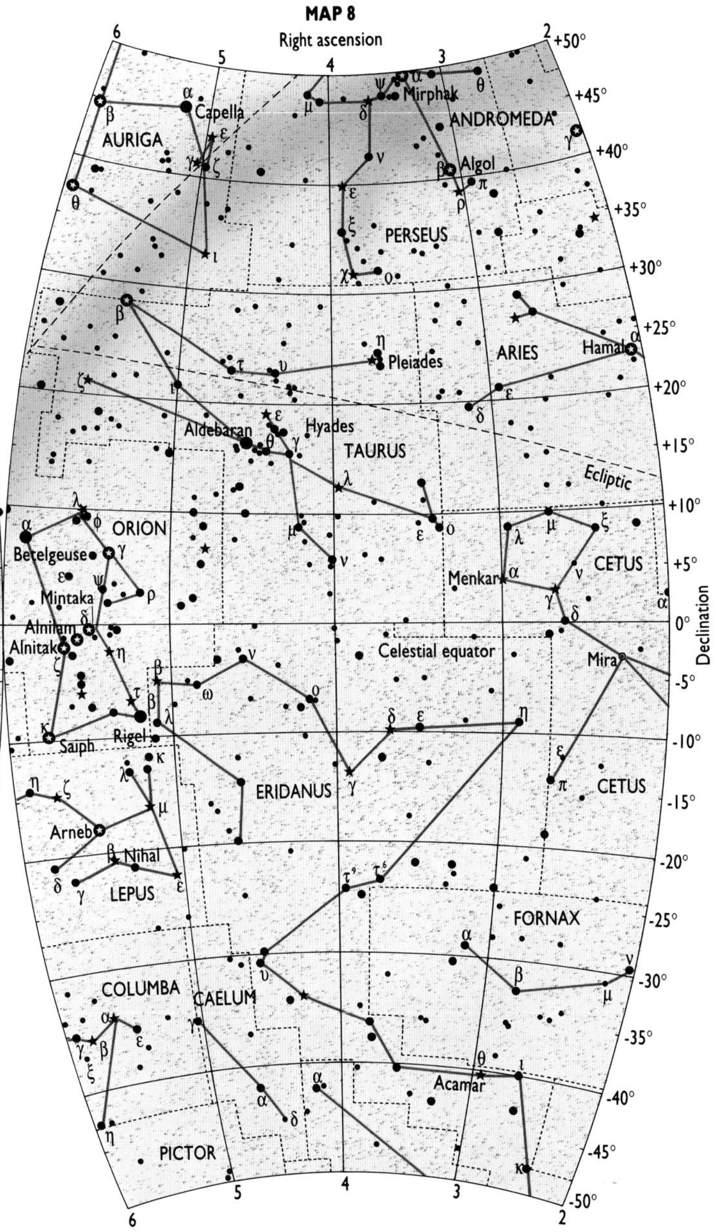
MAP 8
Right ascension
6
5
4
3
2
+50°
+45°
+40°
+35°
+30°
+25°
+20°
+15°
+10°
+5°
0°
-5°
-10°
-15°
-20°
-25°
-30°
-35°
-40°
-45°
-50°
Declination
Capella
AURIGA
Mirphak
ANDROMEDA
Algol
PERSEUS
Pleiades
ARIES
Hamal
Aldebaran
Hyades
TAURUS
Ecliptic
ORION
Betelgeuse
Mintaka
Alnilam
Alnitak
Menkar
CETUS
Celestial equator
Mira
Saiph
Rigel
ERIDANUS
CETUS
Arneb
Nihal
LEPUS
FORNAX
COLUMBA
CAELUM
Acamar
PICTOR

***At magnitude –1.43, Sirius** (α Canis Majoris) is the brightest star in the sky. It is also close to us by astronomical standards – at 8.6 light-years. The Galaxy in Monoceros shows rich star fields through binoculars.*

***Spectacular Orion** includes the Orion nebula below Alnilam, the central star of the belt. Look at the nebula with binoculars and the stars within are seen to be surrounded by a shining gas. It is a region of star birth.*

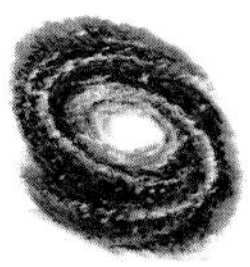

κ kappa λ lambda μ mu ν nu ξ xi ο omicron π pi ρ rho σ sigma τ tau υ upsilon φ phi χ chi ψ psi ω omega

See also

OBSERVATORY EARTH
- Celestial landmarks 16/17
- Viewing the sky 18/19
- Mapping the sky 22/23
- Skymaps 24/29

THE PLANETS
- Wandering stars 52/53

SUN AND STARS
- Star bright 112/113
- Double stars 116/117
- Variable stars 118/119

NEBULAE AND GALAXIES
- Clouds that never rain 132/133
- Star clusters 136/137
- Islands in space 138/139
- Our galaxy 144/145

Phases and reflections

The regular changes in the shape of the Moon and some of the planets reveal much about why they shine and the way they orbit.

The Moon is not the only object in the sky to have phases – other bodies in the sky show them, too. But the waxing and waning of the Moon's disk from crescent to full and back are among the most familiar and obvious sights in the night sky.

Although the Moon seems to give off a great deal of light, it does not actually generate its own illumination as it goes through its phases. What makes the Moon shine is sunlight reflected off the Moon's surface, like a mirror ball in a nightclub lit up by a powerful spotlight.

The Moon shines so brightly that it is hard to believe that it is quite a dark body. The rock and dust of the lunar landscape would look almost black if it were transported to Earth. Yet it reflects

The Moon's phases reflect its progress in its orbit around the Earth (right). It begins as a new Moon which, because it is in direct line of sight with the Sun, cannot be seen from Earth. Next comes a thin crescent, setting just after the Sun in the western twilight. A few days later, it is roughly due south at sunset (due north in the southern hemisphere) and shows half a disk. This is called first quarter, since the Moon is a quarter through its orbit around the Earth.

Seven days after the first quarter, it rises as a full Moon in the eastern sky opposite the Sun at the time the Sun sets. Before and after full Moon, it is called gibbous, which means hump-shaped. Seven days later, the Moon again shows a half disk – at last quarter – but this time it faces the opposite direction, though still with its illuminated side toward the Sun.
A few days after that, it is a thin crescent again, rising before the Sun in the east. These phases are seen worldwide on the same dates.

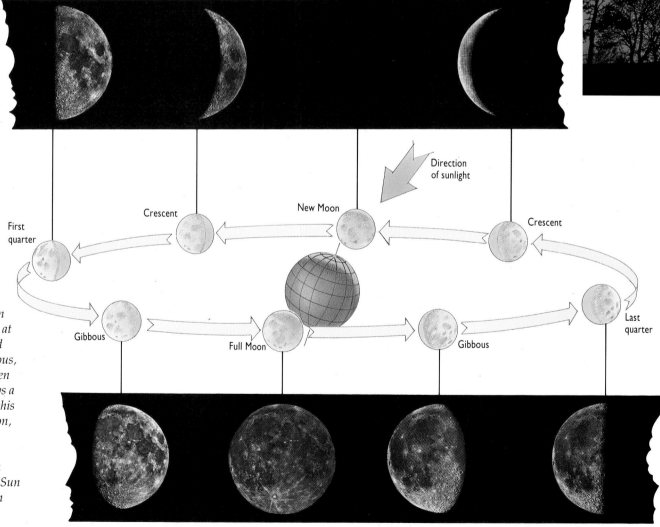

Direction of sunlight

Crescent New Moon Crescent

First quarter

Gibbous Full Moon Gibbous Last quarter

*The crescent Moon is seen only in a twilight sky (**left**), as the Moon must be close to the Sun in line of sight to show this phase. It is rare to see a crescent less than 20 hours after the moment of new Moon.*

Much attention is given to the earliest visibility each month of the crescent Moon, as some calendars are based on it. But since the phases of the Moon are irregular, our own calendar long ago abandoned the attempt to link the months to them.

Illumination on the dark segment of the crescent Moon is caused by earthshine – sunlight reflected from Earth. (If you were standing on the Moon at this time, the Earth would appear almost full.) The brightness of the earthshine on the Moon is partly governed by the amount of cloud cover on Earth. Clouds are more reflective than the Earth's surface, so the more cloud cover on Earth, the brighter the earthshine.

Mercury and Venus both show phases like those of the Moon. They orbit closer to the Sun than to the Earth and appear in full phase only when they are on the opposite side of the Sun from the Earth. At other times during their orbits, they show a range of phases including half phase and even thin crescents.

The phases of Mercury can be seen only through a small telescope. But those of Venus can be seen with binoculars. Mars is the only other planet to show any noticeable phases to observers on Earth, but it is never less than 89 percent illuminated.

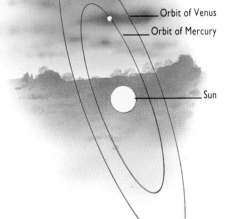

Orbit of Venus

Orbit of Mercury

Sun

Do not search for Mercury or Venus using optical instruments when the Sun is visible. **A glimpse of the Sun through binoculars or a telescope will cause serious eye damage and possibly blindness.**

enough sunlight to look brilliant when it is in the night sky, particularly when full.

The phases of the Moon are caused by the fact that it orbits the Earth while being illuminated by and reflecting light from the Sun. The planets also shine by reflecting sunlight to show phases, though from our viewpoint on Earth, only those of Mercury and Venus can be clearly seen.

The Moon takes 27.3 Earth days to orbit the Earth. During this period, observers on the Moon would experience about two weeks of sunlight and two weeks of darkness. This would happen no matter where on the Moon they were located, since the Moon also takes this time to spin once on its axis.

The interval between successive new Moons as seen from Earth, however, is actually 29.5 days. This is because the Earth is orbiting the Sun, and the changing angle of the sunshine on the Moon, as seen from Earth, slows down the phases.

In the shadow

More than any other science, astronomy depends on light. But there are times when an absence of light can be highly significant and informative.

By complete coincidence, the Moon and Sun appear almost exactly the same size in the sky, so the Sun's bright surface can be blotted out when the Moon's orbit around the Earth takes it between us and our star. This happens surprisingly rarely – only about once every six months – because the plane of the Moon's orbit is at an angle to that of the Earth, and the bodies rarely line up exactly.

Even when the Moon is between the Sun and Earth, you have to be in the right place on Earth to see the Moon's disk cover just part of the Sun, because the Moon is much smaller than the Sun and so casts a cone-shaped shadow. Where the Moon completely covers the Sun, the shadow cast on the Earth is absolute, and observers experience a total eclipse. During totality, as it is known, the sky grows dark, as it is at nightfall, and the brighter stars and planets appear in the sky.

The Earth can also come between the Sun and the Moon. On these occasions, anyone standing on the Earth-facing side of the Moon would see the Sun totally eclipsed by the Earth. From Earth we just see the Moon passing through the Earth's shadow – a total eclipse of the Moon.

Sometimes in such an eclipse, instead of looking completely black, the Moon turns deep red. This happens because some of the Sun's light is bent by Earth's atmosphere to illuminate the Moon's surface. The light is red because red light wavelengths are refracted less. The red color can be very strong, and varies from eclipse to eclipse.

*Creating an eerie daytime darkness on the Earth, the disk of the Moon covers the Sun to form a total eclipse (**above**). Around the dark circle of the Moon glistens the corona, the Sun's outer atmosphere.*

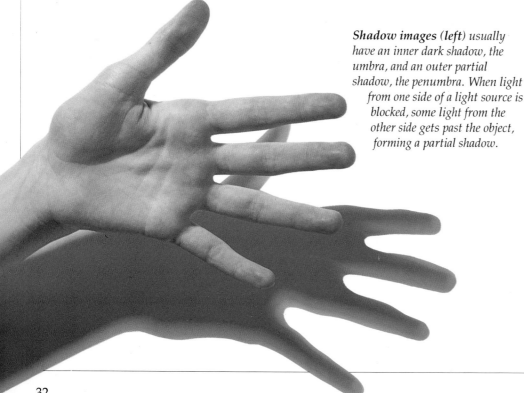

*Shadow images (**left**) usually have an inner dark shadow, the umbra, and an outer partial shadow, the penumbra. When light from one side of a light source is blocked, some light from the other side gets past the object, forming a partial shadow.*

OCCULTATIONS

Occasionally the Moon can move in front of a star or a planet. This event is known as an occultation. These are of interest to astronomers since they provide a means of accurately measuring a body's position in the sky. In 1963 radio astronomers used the method to confirm the positions of the first quasars to be discovered. They timed the instant at which the radio signals from some very small radio sources were cut off by the edge of the Moon.

Even rarer is an occultation of a star by a planet. In some cases, odd effects are noticed. If the planet has an atmosphere, the starlight may be bent around it, just like the sunlight during a total eclipse of the Moon. Previously unknown rings around Uranus were discovered during an occultation when astronomers noticed that light from a star flickered before and after it was obscured by the planet.

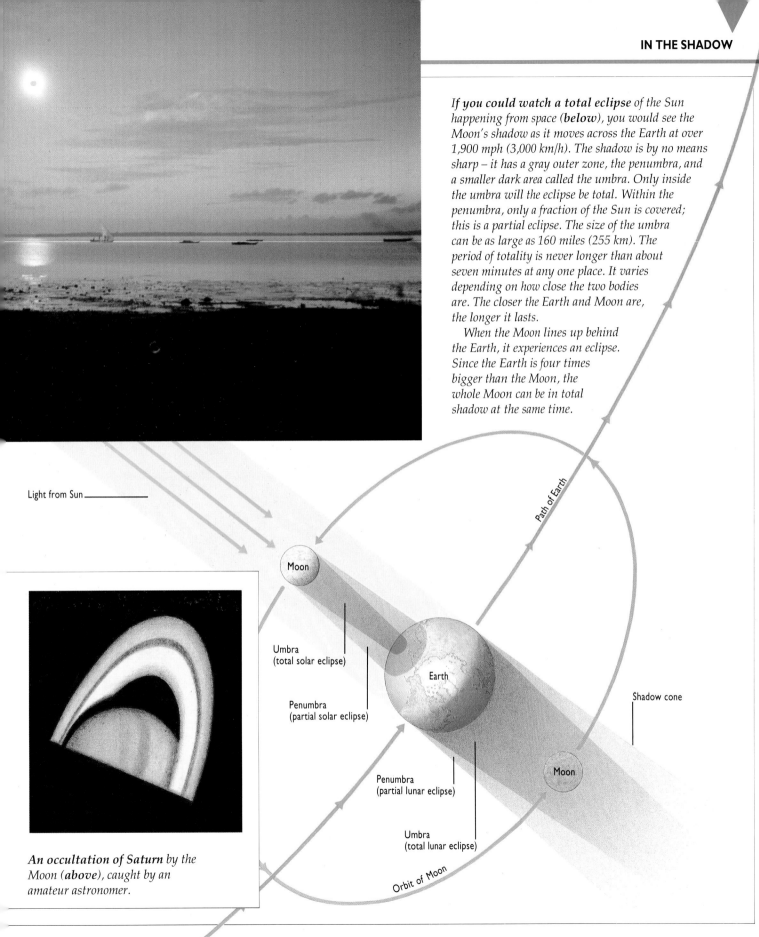

*If **you could watch a total eclipse** of the Sun happening from space (**below**), you would see the Moon's shadow as it moves across the Earth at over 1,900 mph (3,000 km/h). The shadow is by no means sharp – it has a gray outer zone, the penumbra, and a smaller dark area called the umbra. Only inside the umbra will the eclipse be total. Within the penumbra, only a fraction of the Sun is covered; this is a partial eclipse. The size of the umbra can be as large as 160 miles (255 km). The period of totality is never longer than about seven minutes at any one place. It varies depending on how close the two bodies are. The closer the Earth and Moon are, the longer it lasts.*

When the Moon lines up behind the Earth, it experiences an eclipse. Since the Earth is four times bigger than the Moon, the whole Moon can be in total shadow at the same time.

See also

OBSERVATORY EARTH
▶ The home observatory 20/21

▶ Phases and reflections 30/31

THE PLANETS
▶ Companion planet 62/63

▶ The face of the Moon 64/65

▶ Uranus 86/87

SUN AND STARS
▶ A regular star 102/103

▶ The face of the Sun 104/105

▶ Variable stars 118/119

Light from Sun

Path of Earth

Moon

Umbra (total solar eclipse)

Penumbra (partial solar eclipse)

Earth

Penumbra (partial lunar eclipse)

Shadow cone

Moon

Umbra (total lunar eclipse)

Orbit of Moon

An occultation of Saturn by the Moon (**above**), caught by an amateur astronomer.

33

Earth: astronomy's baseline

Simple observations show that the Earth is a globe and provide a surprisingly accurate method of measuring its diameter.

The shape and size of the Earth have been open to speculation throughout history. According to Hindu belief, for example, it is a large disk balanced on the backs of four elephants, who in turn stand on the shell of a giant turtle. Another view was that the Earth was flat.

However, it has been known for about 2,500 years that the Earth is a sphere. The Greek philosopher Aristotle summarized the case for this with two compelling pieces of observational evidence. First, during a total eclipse of the Moon, the Earth's shadow is clearly curved as it encroaches on the Moon. Second, as you travel north or south and move around the Earth's curve, different stars appear above the horizon, and some stars become visible in the south that are never seen farther north.

Some Greek mathematicians even estimated the size of the Earth, and by 200 B.C. they got it approximately correct. But perhaps the most famous miscalculation of the Earth's size occurred when Christopher Columbus estimated that the coast of Asia would be found only 3,900 miles (6,275 km) west of the Canary Islands. Instead of India or China, he found the Americas. Today radar mapping and artificial satellite images have allowed us to make accurate measurements of the size of the Earth and the distances across its surface. Once the size of the Earth was known, astronomers could take the next step in measuring and work out the distance to the Moon using trigonometry and two distant points on the Earth's surface as a baseline.

The Earth's curvature is revealed by the fact that the horizon is farther away when you climb up a tall object such as a skyscraper (right). You are, in fact, seeing around the curvature of the Earth. At ground level (below), with your eyes at a height of 6½ feet (2 m) above sea level (b), the horizon (b) is 3½ miles (5.5 km) away.

From the top of a 1,300-foot (400-m) skyscraper (a), the horizon is 47 miles (75 km) away (a), and from the summit of a 10,000-foot (3,000-m) mountain, it would be over 125 miles (200 km) away.

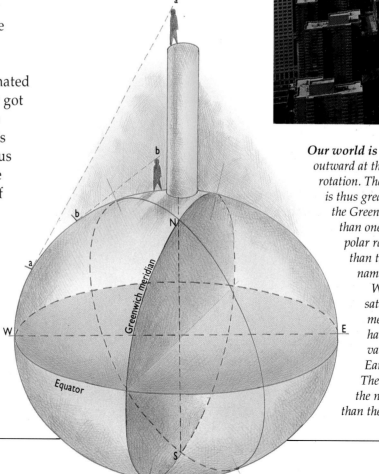

Our world is not a perfect sphere – it bulges outward at the equator (left) as a result of its rotation. The circumference around the equator is thus greater than the circumference around the Greenwich meridian. The bulge is less than one part in 300; in other words, the polar radius is 13 miles (21 km) shorter than the equatorial radius. The technical name for this shape is an oblate spheroid.

With the launch of Earth-orbiting satellites, extremely accurate measurements became possible. These have shown that there are additional variations to this spheroid, and the Earth is actually slightly pear-shaped. The distance from the Earth's center to the north pole is 148 feet (45 m) longer than the distance to the south pole.

ERATOSTHENES AND THE EARTH

The first true measurement of the Earth's diameter was made by the Greek map maker, Eratosthenes, in about 200 B.C. He was aware that in the Egyptian town of Syene, no shadow is cast in a well on midsummer's day – the Sun is directly overhead. At Alexandria, however, an upright pole has a shadow on this day. From the shadow's length, he was able to calculate the angle of the Sun from the vertical. This angle of 7.2 degrees equals the latitude difference between the towns.

Knowing the distance between the two places, he worked out the Earth's circumference to be 25,200 miles (40,555 km), only 1 percent more than today's accepted figure of 24,900 miles (40,074 km).

7.2°

Light from Sun

Shadow cast in Alexandria on midsummer's day

Distance between Alexandria and Syene

Sun shines vertically into well in Syene on midsummer's day

7.2°

Since the first attempts to calculate the circumference of the Earth, its diameter has been used as a starting block for measuring systems. For instance, in 1790, following the revolution of 1789, the French government replaced the old imperial system of measuring with a more logical decimalized one. They decided upon a measurement that was one ten-millionth of the distance from the equator to the pole: the meter.

The length for the meter that they calculated, based on their knowledge of the Earth's size, was slightly inaccurate. The meter is now set as the distance traveled by light in a vacuum in one 299,792,458th of a second.

Measuring the system

Simple formulae describe the motions of the planets and can be used to work out the size of the solar system.

Once the diameter of the Earth was known, astronomers had a means of measuring the distance to the Moon. They used the method of trigonometric parallax – this works on the same principle as judging distances using the stereoscopic vision of both our eyes. When an object is seen from two different viewpoints, it appears to shift against the background. By observing the Moon from two widely spaced places on Earth whose separation is known, it is seen against different background stars. The angles between the different background stars are measured; then, by using basic trigonometry, it is a simple task to calculate the distance between us and the Moon.

The most promising method to find the distances to the known planets was to time the passage, or transit, of Venus across the face of the Sun from two different places at opposite ends of the Earth. The difference in viewing position would give two slightly differing times for the start of the transit across the Sun. And from these two different times, a parallax angle can be calculated giving a distance to Venus, which comes closer to Earth than any other planet in the solar system. The distances to the other planets can be worked out by slotting the figure for Venus into the planetary motion equations of Johannes Kepler. Transit observations involved expeditions to remote regions of the Earth where the transits could be observed and quite accurate figures were calculated.

Today trigonometric parallax has given way to more sophisticated methods involving the reflection of radar and laser pulses. These give astronomers far more precise results.

Modern techniques allow astronomers to calculate the distances from Earth to the Moon and the planets with great accuracy. Radar pulses and light pulses from lasers (right) can be aimed at the objects. Because light and radar pulses travel at a known velocity – the speed of light, about 186,300 miles/s (300,000 km/s) – the distances can be calculated from the time taken to get there and back.

Reflectors left on the Moon by the Apollo missions have proved useful in providing targets at which astronomers can aim their lasers. Although the intensity of the laser beam is extremely low (no more than a dribble of photons) by the time it gets back, it nevertheless allows the distance between the Earth and the Moon to be calculated to within a couple of centimeters.

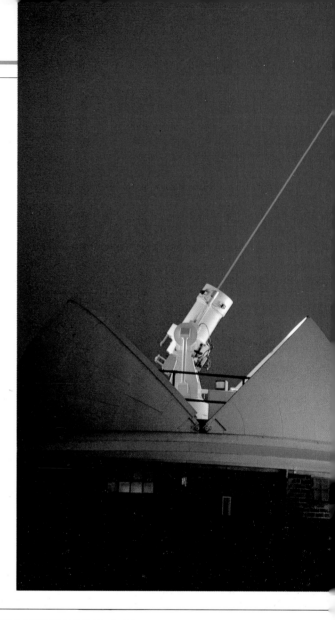

JOHANNES KEPLER – DEFINING THE PATHS OF THE PLANETS

In 1609, the year that Galileo first turned his telescope on the heavens, Johannes Kepler (1571–1630) began to publish his laws of planetary motion for Mars. By 1621 Galileo's observations and Kepler's laws finally discredited the theory that the Sun and planets moved around the Earth. Kepler used data gathered by the Danish observer Tycho Brahe (1546–1601) to try to prove this geocentric view of the universe.

Tycho had measured the positions of the planets for 20 years to the highest accuracy possible without the use of a telescope. Kepler's task was to find orbits that would best fit Tycho's observations. He had great problems with the system propounded by Copernicus, in which the planets, including the Earth, moved in circles around the Sun. Eventually Kepler found that the simplest way to account for the uneven movements of the planets, and Mars in particular, was to assume that they moved in ellipses rather than circles, with the Sun at one focus of the ellipse rather than at the center.

Before the advent of radar and laser measuring equipment, the transit of Venus across the face of the Sun was important in calculating the size of the solar system. In 1768 Captain James Cook (**left**) set off on an expedition to the South Seas with the intention of accurately timing the onset of Venus's transit, due to occur in 1769. It was on this voyage that he first sighted Australia.

Background stars

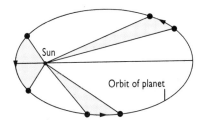

Moon

Parallax angle

Earth

The relatively simple mathematics of trigonometric parallax is used to calculate the huge distances across our solar system and even out to the nearest stars. The basis for the calculation is the apparent movement of a body in degrees against the background when viewed from two ends of a baseline of known length. Map makers use the same method on Earth by measuring the angles from the ends of a known baseline to a distant object using a theodolite. For nearby objects, such as the Moon and the closer planets, this baseline needs only to be as long as the diameter of the Earth to produce a sufficient apparent movement (**left**). From the parallax angle, measured at opposite ends of the baseline, the astronomer has enough information to work out the distance to the object using simple trigonometry.

For objects beyond the solar system, a larger baseline is needed, requiring each end of the Earth's orbit to create a measurable apparent movement against the background stars.

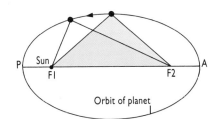

P — Sun — F1 — F2 — A

Orbit of planet

Kepler's first law (above) shows that the orbit of a planet is an ellipse, not a circle, with the Sun at one of the two foci (**F1** and **F2**). The planet is at its closest to the Sun at the perihelion (**P**) and farthest away at aphelion (**A**). The sum of the distances from the planet to the two foci is always the same.

Sun

Orbit of planet

The second of Kepler's laws (above) describes how the area covered by a line between the planet and the Sun, namely the radius vector, is equal for identical periods of time anywhere in the planet's orbit. This is because the planet accelerates as it nears its perihelion and slows down as it moves away.

Kepler's third and final law of planetary motion (**below**) shows how orbital velocity decreases as distance from the Sun increases. Orbital velocity is inversely proportional to the orbital radius – the square of the length of a planet's year varies according to the cube of its distance from the Sun.

This being so, astronomers need only to find the orbital velocity of a planet to discover its distance from the Sun. Observations of the solar system show that the Earth travels at 18½ miles/s (29.8 km/s), while Saturn travels around the Sun at 6 miles/s (9.6 km/s).

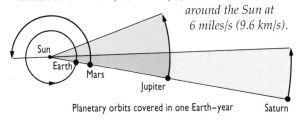

Sun
Earth
Mars
Jupiter
Saturn

Planetary orbits covered in one Earth-year

Light-years

Stars are so far away that time is used as a measure of their distance.

For centuries astronomers have tried to calculate the distances to the planets and stars we see in the night sky. Since we cannot actually travel to the stars, this can be done using trigonometric parallax (see opposite). For objects close to Earth, such as the Moon and nearer planets, astronomers can take opposite sides of the Earth as the two ends of the baseline. But this is too short for distant bodies like the stars. So efforts were made to see whether their positions varied when they were observed from opposite sides of the Earth's orbit – giving a baseline nearly 186 million miles (300 million km) long.

In 1838 German astronomer F.W. Bessel used this method to measure the parallax of a star – 61 Cygni. He found its annual parallax to be one-third of an arc second (about one ten-thousandth of a degree) and thus calculated its distance from Earth to be over 62 trillion miles (100 trillion km). But numbers like these are too big for everyday use, so instead astronomers work in light-years or parsecs.

A light-year is based on the distance light travels in a single year. The speed of light is constant – it is always approximately 186,300 miles/s (300,000 km/s) – so light moves about 6 trillion miles (10 trillion km) during the course of a year. It takes 11.2 years for light from 61 Cygni to reach us, so the star is 11.2 light-years away. One parsec (pc), a unit which is widely used by astronomers because it is linked to parallax measurement, equals 3.26 light-years. So 61 Cygni's distance from Earth in parsecs is 3.43.

1,000 years

100 years

10 years

⇦ Time

1 year

The times at which light set off from objects in the past depend on how far away the objects are. Around 1,000 years ago, in the time of the Vikings, light we now see from Rigel was just about to leave the star.

Light from Proxima Centauri set out 4.3 years ago, when a child now aged 10 was nearly six.

Traveling from Pluto at the outer limits of the solar system, light takes on average 5.4 hours to reach us. It will have begun its journey as the child started school for the day.

10 hours

1 hour

In one minute light travels 11 million miles (18 million km), the equivalent of 450 times the Earth's circumference. Light from the Sun – about 93 million miles (150 million km) away – thus takes 8.3 minutes to reach Earth. It will have begun its journey as this 10-year-old child was on her way home from school.

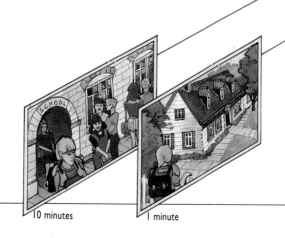

10 minutes

1 minute

In just one second light travels nearly 186,300 miles (300,000 km). Light covers the distance between the Earth and the Moon, for instance, in just over one second. Its journey will have started as the child began to blow the seeds off the dandelion.

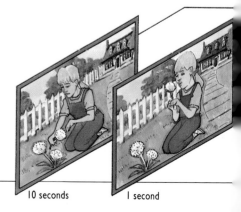

10 seconds

1 second

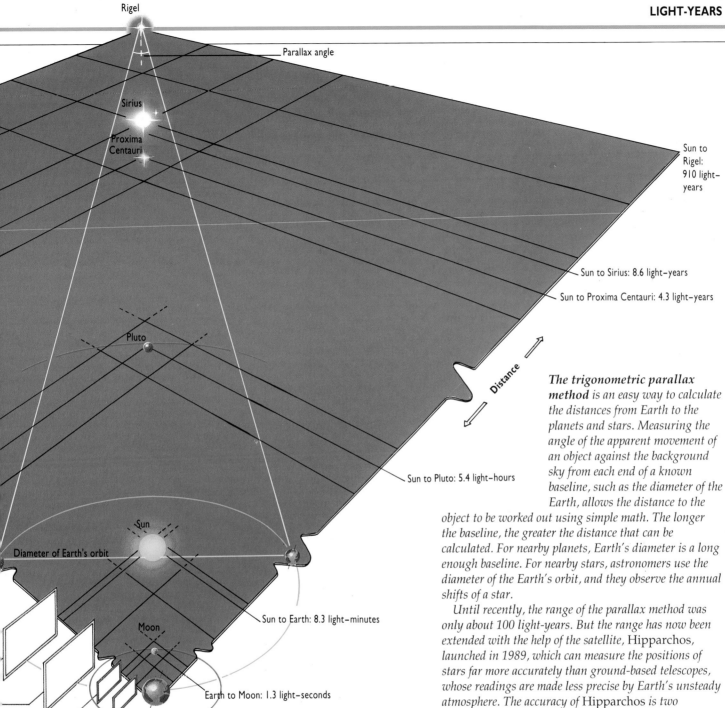

Rigel

Parallax angle

Sirius

Proxima Centauri

Sun to Rigel: 910 light-years

Sun to Sirius: 8.6 light-years

Sun to Proxima Centauri: 4.3 light-years

Pluto

Distance

Sun to Pluto: 5.4 light-hours

Sun

Diameter of Earth's orbit

Sun to Earth: 8.3 light-minutes

Moon

Earth to Moon: 1.3 light-seconds

Earth

Now

The trigonometric parallax method *is an easy way to calculate the distances from Earth to the planets and stars. Measuring the angle of the apparent movement of an object against the background sky from each end of a known baseline, such as the diameter of the Earth, allows the distance to the object to be worked out using simple math. The longer the baseline, the greater the distance that can be calculated. For nearby planets, Earth's diameter is a long enough baseline. For nearby stars, astronomers use the diameter of the Earth's orbit, and they observe the annual shifts of a star.*

Until recently, the range of the parallax method was only about 100 light-years. But the range has now been extended with the help of the satellite, Hipparchos, launched in 1989, which can measure the positions of stars far more accurately than ground-based telescopes, whose readings are made less precise by Earth's unsteady atmosphere. The accuracy of Hipparchos is two thousandths of an arc second, 30,000 times more acute than the naked eye. The satellite has been used to provide the distances of stars out to around 1,000 light-years.

One of the farthest stars that can have its distance measured using parallax is Rigel (Beta Orionis), which is about 900 light-years away. Closer, at 8.6 light-years, is the brightest star in the sky, Sirius (Alpha Canis Majoris). The closest star to Earth is Proxima Centauri which, at 4.3 light-years away, has an annual parallax angle of just under 1 arc second (three ten-thousandths of a degree).

Light from nearby objects reaches the eye so quickly that they appear as they are now – that is, as the 10-year-old child blows the dandelion's seeds. What we see to be happening in the universe "now" is actually made up of light that set off some time in the past.

Deep space

Light from remote objects helps plumb the vast distances across space.

There are many ways to measure the distances to objects across the universe. For close stars, it can be done directly using simple trigonometric calculations based on measurements of a star's parallax, or annual movement against the stellar background. Beyond the 1,000 or so light-years that is the limit for trigonometric parallax, however, astronomers have to use indirect means of measurement. These allow them to determine the distances to remote objects millions and even billions of light-years away.

Most indirect methods use what are called "standard candles" – objects whose real brightnesses are known. If an object's absolute magnitude (luminosity) is known, then its distance can be calculated by comparing its absolute magnitude with how bright it seems from Earth (its apparent magnitude).

Indirect measuring techniques still use trigonometric parallax as their starting point. For instance, using this method to figure out the distance of nearby stars allows their absolute magnitudes to be calculated. From this, it has been found that for most stars there is a direct relationship between their color and their absolute magnitude. So a star of a particular hue has a predictable absolute magnitude. The absolute magnitude can then be compared with its apparent magnitude to give its distance. This technique, known as main sequence fitting, is effective out to about 30,000 light-years.

Cepheids are some of the most useful standard candles. They are easy to recognize because they vary their light output in a regular way and can be seen in other galaxies. The period of variability, from one maximum to another, is related to the absolute magnitude. So once a Cepheid variable is found, and its period measured, astronomers know its absolute magnitude and can then figure out its distance from Earth.

Other objects that have predictable absolute magnitudes include planetary nebulae (the ring-shaped gas clouds resulting from a Sun-sized star's death throes); globular clusters (the groups of stars that accompany galaxies); rotating galaxies; and certain supernovae. Standard candles are used as distance markers. If a supernova is spotted in a galaxy, for instance, the distance of the supernova gives the galaxy's distance as well.

On a bridge stretching into the distance (above), streetlights with equal light output appear dimmer the farther away they are. There is, in fact, a simple relationship between brightness and distance – the brightness of an object is inversely proportional to the square of its distance. So if one light is twice as far away as another of identical brightness, it appears four times fainter.

This simple principle is important in astronomy. Many methods employed to calculate distance compare the apparent brightnesses of objects in the night sky with their true brightnesses.

Distance measuring is taken step by step to the universe's edge by a variety of methods. Underlying almost all of them is the general principle that if the absolute magnitude of an object is known or can be worked out, then comparing it with its apparent magnitude – how bright it appears from Earth – allows a simple calculation to give its distance. In the diagram (**below**) the bars of color mark the distance over which a particular measuring technique can be used. The technique is identified in the small image which has a frame the same color as the bar to which it applies. The scale of distances is on the right of the diagram. Thus measurements of distance based on the use of the predictable absolute magnitudes of Type 1 supernovae (yellow bar and yellow frame) can be made between about 100,000 and 320 million light-years. And planetary nebulae (orange bar and frame) can be seen as far away as 50 million light-years.

The distances to extremely remote objects, such as quasars and distant galaxies, are assessed by observing the red shift of lines in their spectra. The greater the red shift, the more distant the object.

Light now arriving at Earth from distant objects set off in the past – the farther away, the farther back in time. For instance, light from a star 10,000 light-years away set off when farming was invented on Earth. Light from an object 100 million light-years away set off during the age of the dinosaurs.

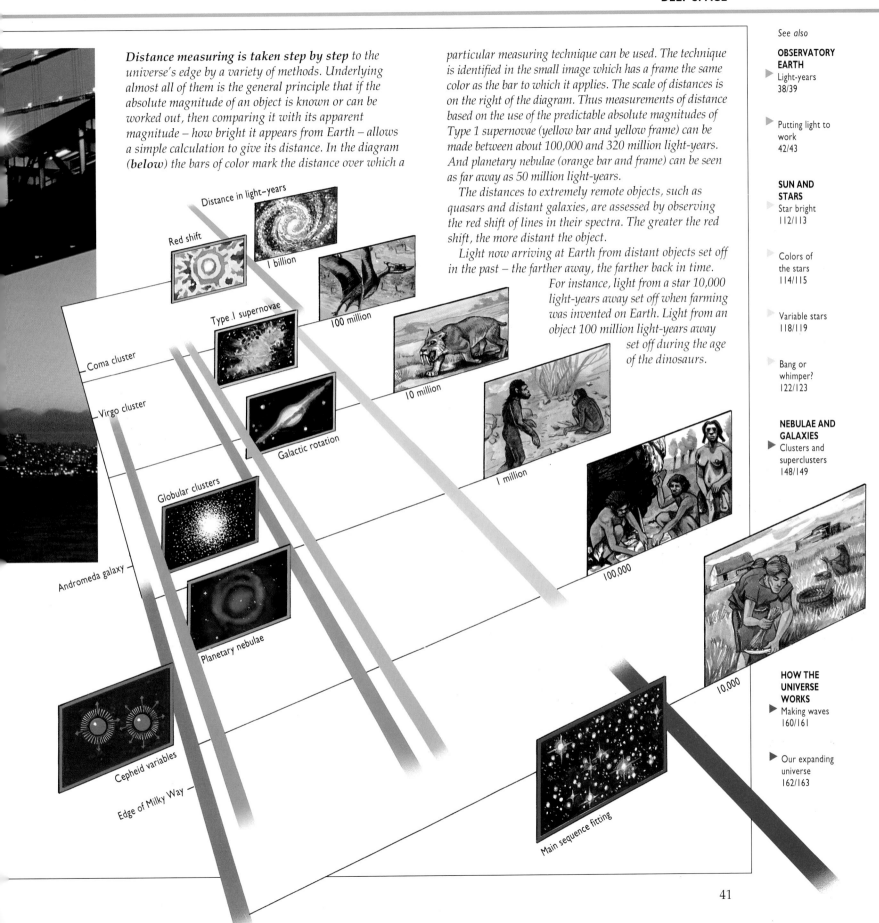

Distance in light–years

Red shift

1 billion

Type I supernovae

100 million

Coma cluster

Virgo cluster

10 million

Galactic rotation

1 million

Globular clusters

Andromeda galaxy

100,000

Planetary nebulae

10,000

Cepheid variables

Edge of Milky Way

Main sequence fitting

Putting light to work

Light can tell us much about its source and the medium it passes through on its way to Earth.

What we see as white light is actually the combined effect of all the colors of the rainbow acting together. It is the visual equivalent of hearing all the instruments of an orchestra playing all the notes at once. Even though piccolos and tubas make quite dissimilar sounds by themselves, when they are played together the effect is different. When we see a rainbow, we are seeing a spectrum – light spread out according to its wavelength. Blue light has a short wavelength (just as the high pitch of a piccolo has a short wavelength), while red has a longer wavelength (equivalent to the tuba's lower pitch).

Astronomers study light by passing it through a narrow slit and then splitting it up. This allows them to see a spectrum made up of an image of the slit repeated at many wavelengths. Each source of light has a characteristic spectrum. The spectrum of a light bulb, for example, is a pure rainbow of color. This is the basic spectrum of an incandescent body. Other bodies, such as the glowing gases in nebulae, produce an emission spectrum. This consists simply of bright lines of specific colors, which are caused by atoms within the body emitting photons whose wavelengths match those of the bright lines. Each element creates its own characteristic spectrum with its own pattern of lines.

When light from an incandescent body such as a star has passed through low-temperature gas, an absorption spectrum is seen. Dark lines appear, where certain colors are missing. The "absent" wavelengths have been absorbed both by gases in the outer layers of the star, and by gases in space between the star and Earth. Dark lines in a spectrum can tell astronomers about gases present in the star and in space.

White light splits into colors when it passes through a prism (above) because each wavelength is refracted (bent) by a different amount on its way through. The result is the splitting of visible light into its component colors of violet, indigo, blue, green, yellow, orange, and red. The wavelengths of light range from that of violet, which is about 380 nm, to that of red, which is 750 nm.

WAVES OF LIGHT

Electromagnetic radiation, of which visible light is just one type, is made up of both electric and magnetic fields. The fields lie at right angles to each other and to the direction in which the radiation travels. The wavelength of the radiation is the distance between two consecutive wave peaks.

Each wavelength corresponds to a specific color. Helium atoms, for example, can emit light with a wavelength of 587.6 nm (a nanometer is one billionth of a meter), causing a yellow line in a spectrum (**above**).

Emission spectrum of helium

587.6 nm

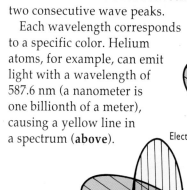

Electric field

Wavelength

Magnetic field

Radiation direction

Sodium emission line

Emission spectrum of streetlight

The spectrum of a sodium streetlight (above) shows the yellow line that is the hallmark of the element. It is created by excited sodium atoms emitting light at a wavelength of about 589 nm, in the yellow part of the spectrum.

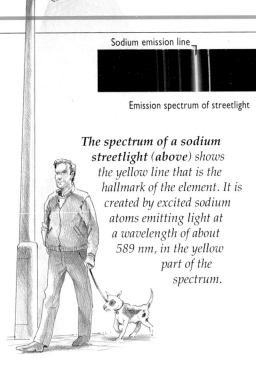

When white light passes through a stained glass window (right), *each different type of glass absorbs particular colors from pure white light. Red glass, for instance, absorbs all colors except red. In a similar way, a gas absorbs particular wavelengths from light passing through it to form a spectrum with lines of color missing, called an absorption spectrum.*

The image on a TV screen (above) has similarities to the emission spectrum from a glowing gas. The bright dots on the screen are excited by being bombarded by electrons. When excited, the phosphors of which they are made emit electromagnetic radiation, and the three primary colors of the light (red, blue, and green) produced by the dots combine to give a picture on the screen.

The spectrum created by the Sun contains a large number of absorption lines. These are the result of photons of a specific energy (and therefore wavelength) being absorbed by elements in the Sun's atmosphere. There are about 20,000 lines in total, of which only the strongest are shown here (right). So far, over 70 elements in the Sun have been identified from lines in its spectrum, the most common of which is hydrogen.

Solar spectrum with absorption lines

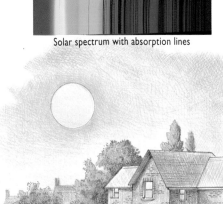

To the end of the dial

Astronomers use energy invisible to the eye to probe deep into the heavens and extend our knowledge of the universe.

The spectrum of visible light, the familiar rainbow, is only a tiny part of the full range of the spectrum of energy that is electromagnetic radiation. To astronomers the wavelengths extending far on each side of red and blue are as vital as light.

Beyond the red end of the spectrum, the wavelength of the radiation gets longer and its energy progressively less. The first type of radiation is infrared, which has wavelengths between 0.00075 mm and 1 mm. We can detect infrared radiation with our skin, as warmth from sunshine, for instance.

As the wavelengths get longer, the radiation is called millimeter wave and then microwave radiation, which has wavelengths up to 30 mm long. Next are radio waves with wavelengths from 30 mm up to several tens of kilometers.

Infrared, millimeter, and microwave radiation are produced in the same way as light, by hot bodies such as stars. Some radiation in the microwave part of the spectrum comes not from individual bodies, but from the universe as a whole. This background radiation is a relic of the Big Bang in which it originated. Radio waves, too, are given off by some hot objects

in space. But there is another source of radio radiation, known as synchrotron radiation, which is more powerful. This is made by electrons moving in a magnetic field, as happens in active galaxies. Hydrogen and other molecules in galaxies also give off characteristic radio emissions.

There is a great deal of infrared and microwave radiation coming from space, but much of it is blocked by Earth's atmosphere, so astronomers make their observations either from high mountains or from spacecraft.

Most radio waves make it down to the surface, so radio-telescopes are ground-based. A radiotelescope works in a similar way to a reflecting telescope. Instead of a mirror, there

*Infrared images reveal matter such as the gas and dust surrounding areas of star birth. At infrared wavelengths, such regions appear in the Andromeda galaxy, or M31 (**left**), as a bright yellow ring of cloud. The galaxy's nucleus is also bright due to infrared emitted by dust and other material ejected from old stars that have shed their outer layers. This image was made using observations from the Infrared Astronomical Satellite (IRAS).*

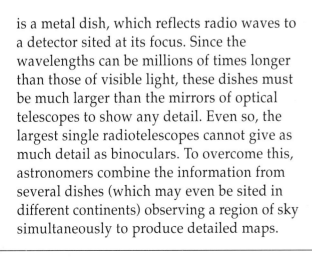

Hydrogen gas gives off radio signals with a wavelength of 21 cm. By mapping these signals, radio astronomers have learned much about the structure of our own and other galaxies, since cool hydrogen gas clouds make up much of the matter between a galaxy's stars.

In this image of the Andromeda galaxy at the 21 cm wavelength (**left**), the amount of hydrogen is shown. The colors are artificially produced to match slight changes in the frequency of the radiation caused by the Doppler effect. Thus, hydrogen approaching the Earth is in blue and green and that receding is in yellow and red. This shows that the gas in the galaxy (which we see edge on) is spinning relative to the Earth.

In visible light wavelengths, the Andromeda galaxy (*above*) shows a great deal of detail, but most of what can be seen is the light coming from hot objects such as the surface layers of stars. Much of the detail of the galaxy is obscured by clouds of gas and dust which are visible at certain radio and infrared wavelengths.

is a metal dish, which reflects radio waves to a detector sited at its focus. Since the wavelengths can be millions of times longer than those of visible light, these dishes must be much larger than the mirrors of optical telescopes to show any detail. Even so, the largest single radiotelescopes cannot give as much detail as binoculars. To overcome this, astronomers combine the information from several dishes (which may even be sited in different continents) observing a region of sky simultaneously to produce detailed maps.

JODRELL BANK – A RADIO EYE ON THE SKY

After World War II, a pioneering radio astronomy observatory was set up at Jodrell Bank, Cheshire, England, by the University of Manchester. It was run by a team led by Bernard Lovell (**right**). The most famous radiotelescope on the site is the 250-foot (75-m) dish (**below**). The observatory also houses MERLIN, or Multi-Element Radio-Linked Interferometry Network. Simultaneous observations made by radiotelescopes across England and Wales are sent to Jodrell Bank, where the data is combined to form the equivalent of a telescope 135 miles (218 km) across.

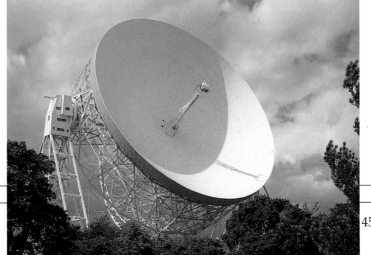

See also

OBSERVATORY EARTH
▶ Viewing the sky 18/19

▶ Measuring the system 36/37

▶ Putting light to work 42/43

▶ High-energy astronomy 46/47

▶ Scale of the universe 48/49

NEBULAE AND GALAXIES
▶ Active galaxies 142/143

HOW THE UNIVERSE WORKS
▶ Making waves 160/161

▶ Echoes of the Bang 176/177

High-energy astronomy

Some of the most dramatic events in the universe are seen only by spacecraft able to detect ultraviolet, X-rays, and gamma rays.

Beyond the blue end of the spectrum lie increasingly energetic radiations with proportionately shorter wavelengths. First is ultraviolet (UV), followed by X-rays, and then gamma rays as the wavelengths shorten. Such high-energy radiation – given off by hot and energetic sources – is blocked by Earth's atmosphere, however, so astronomers can only pick up shorter wavelength radiations using satellite-borne detecting equipment.

Until the space age, astronomers had no way of observing very high-energy radiations. But now that satellites can be put in orbit above Earth's atmosphere, a variety of craft, each with instruments designed to observe in a specific area of the spectrum, have provided data.

*For instance, an experiment called Astro-1 (**left below**), which was carried aboard space shuttle Columbia in 1990, contained the first equipment to make images of objects in the sky in short-wave UV.*

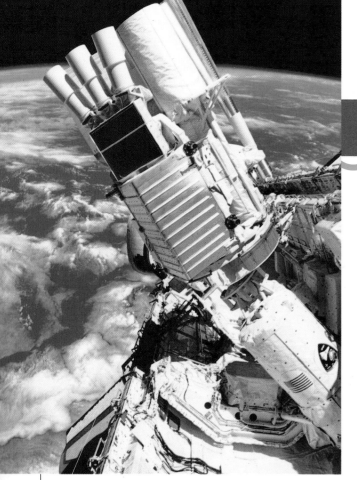

On Earth, UV from our Sun can burn the skin. In general, UV is produced by hot gas – either in stars themselves, the coronas surrounding stars, or the highly active cores of galaxies and quasars. Since its launch in 1978, the International Ultraviolet Explorer (IUE) satellite has observed some 90,000 objects, from planets to quasars. Cold interstellar gas and dust absorbs certain UV wavelengths. By seeing which wavelengths from UV sources have been absorbed by intervening material, astronomers have learned much about this type of matter.

*Our Sun gives off not only light, but also other wavelengths of radiation, including X-rays (**above**). It is not just the well-behaved warmer of our world, it is also a cauldron of violent nuclear reactions. In its core, fusion releases gamma rays and X-rays, which are gradually converted to less destructive and energetic wavelengths of radiation as they move to the Sun's outer layers.*

The X-rays in this image come from the Sun's corona and outer atmosphere where gases are at high temperatures, higher, in fact, than on the Sun's surface.

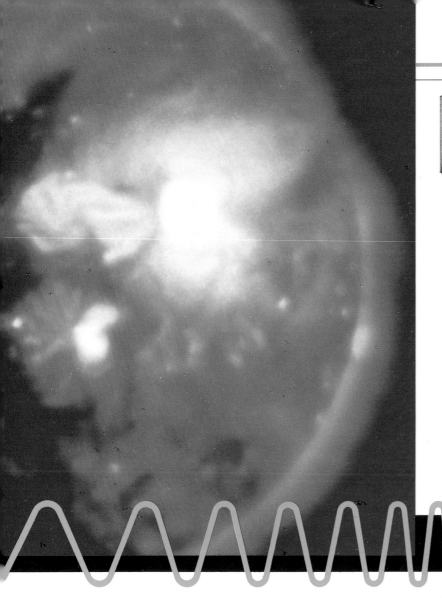

X-rays are invaluable on Earth because they penetrate skin and can be used to make images of internal structures such as bones and some organs.

In the universe at large, they are produced in such locations as active galactic nuclei (AGNs), Seyfert galaxies, and quasars. These are all types of galaxy with bright centers, and astronomers are keen to find out the cause of their outpourings of energy. X-ray observations suggest that the emissions are the result of material falling into giant black holes at the galaxies' centers.

At X-ray wavelengths, the sky is covered with what appear to be stars. These include remote quasars and objects within our own galaxy, such as double stars with white dwarfs as one of their parts. A giant space X-ray observatory, called AXAF, is due to be launched around the year 2000 to study X-ray sources further.

Shortening wavelength, increasing energy

The Tarantula nebula in the Large Magellanic Cloud (one of the closest galaxies to our own Milky Way) shows up as wispy gas filaments in visible light (**below**). The gas is glowing in the light of new stars that are emitting much high-eneregy UV radiation.

The region is also the source of gamma-ray emissions. These rays are produced by the highly energetic bodies left behind after supernovae explosions. These bodies are either super-dense neutron stars or black holes that remorselessly devour matter and light.

Gamma rays are produced by violent, energetic sources, such as supernovae, pulsars, and quasars, and possibly when material is sucked into a black hole. The gamma-ray sky shows a general band of radiation from the line of our galaxy, as well as individual point sources from major supernovae remnants. There are also powerful sources, called gamma-ray bursters, which mysteriously appear, then vanish after a few seconds. For this short time, they can be stronger than all the other gamma-ray sources in the sky together. They seem to come from beyond the Galaxy and so must be generated by extremely powerful events.

47

Scale of the universe

The almost inconceivable size of the universe can be understood with the help of comparisons with more familiar objects.

Across the night sky, a large band of light stretches through the blackness. This is the Milky Way, a giant galaxy of which our sun and solar system form just a tiny part. Using telescopes, astronomers can see that the Milky Way consists of individual stars, yet none of them is discernible separately with the naked eye within this dense band of light. While they appear to blend together in the Milky Way, the stars that make up our galaxy are, in fact, separated by vast distances. For instance, traveling by jumbo jet at about 600 mph (970 km/h) to the nearest star, Proxima Centauri, would take 5 million years, and it would take a further 100 billion years to reach the far side of our galaxy. Times and distances like these are well beyond those that we can directly experience.

Yet increasingly faster modes of getting around have changed our perception of distance. In earlier days, people used to walk. Now we have trains, cars, aircraft, and even spacecraft, all of which have "shrunk" distances.

For the present, the limit of human travel has been the Moon. Beyond this direct experience, we rely upon the information from unmanned space probes and from light and the other forms of radiation we can detect. Our understanding of distance and time also depends on the theories of scientists such as Newton and Einstein who have described the universe in an array of principles, mathematical formulae, and theories. Astronomers use these to probe farther across the universe, and deeper into its history, almost to the point of its cataclysmic birth – the Big Bang.

The diameter of the Sun is 860,000 miles (1.4 million km). At this scale it is 0.019 mm, equivalent to the size of a single-celled protozoan.

The distance from the Earth to the Sun is 93 million miles (150 million km). At this scale it is 2 mm, equivalent to the size of a rice grain.

The diameter of Venus is similar to that of Earth.

The closest approach distance from Earth to Venus is 26 million miles (41 million km). At this scale it is 0.56 mm, equivalent to the thickness of a credit card.

The diameter of the Earth is 7,926 miles (12,756 km). At this scale it is 0.00017 mm, equivalent to the size of a virus.

The distance from the Earth to the Moon is 238,855 miles (384,400 km). At this scale it is 0.0053 mm, equivalent to the size of a human white blood cell.

Scale: Diameter of Milky Way 100,000 light-years = Diameter of Earth 7,926 miles (12,756 km)

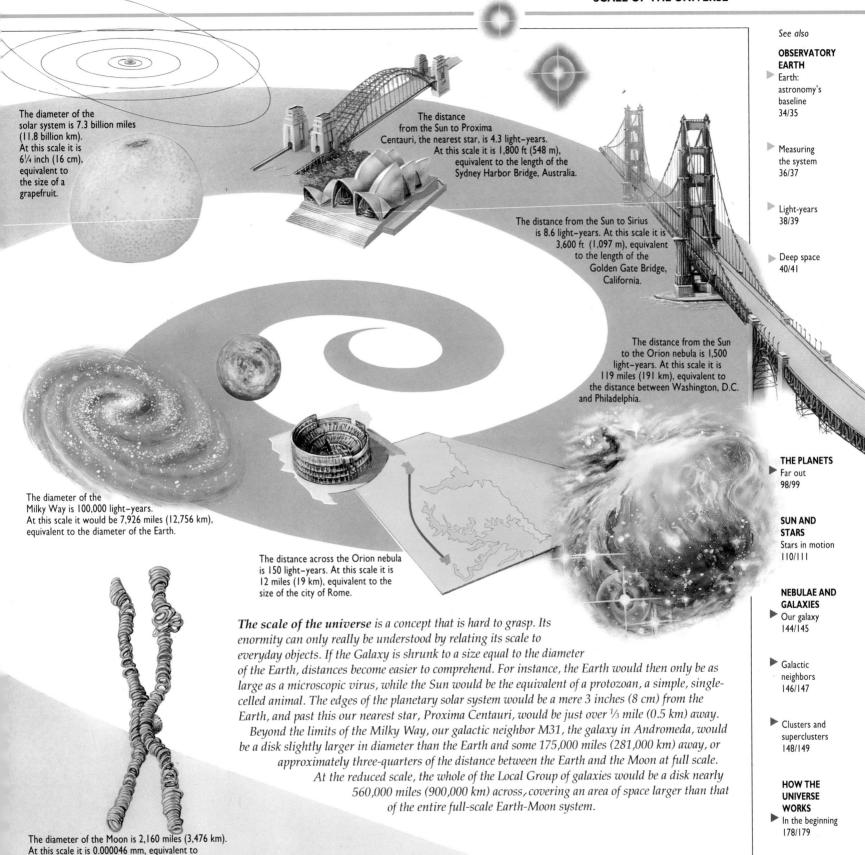

The diameter of the solar system is 7.3 billion miles (11.8 billion km). At this scale it is 6¼ inch (16 cm), equivalent to the size of a grapefruit.

The distance from the Sun to Proxima Centauri, the nearest star, is 4.3 light-years. At this scale it is 1,800 ft (548 m), equivalent to the length of the Sydney Harbor Bridge, Australia.

The distance from the Sun to Sirius is 8.6 light-years. At this scale it is 3,600 ft (1,097 m), equivalent to the length of the Golden Gate Bridge, California.

The distance from the Sun to the Orion nebula is 1,500 light-years. At this scale it is 119 miles (191 km), equivalent to the distance between Washington, D.C. and Philadelphia.

The diameter of the Milky Way is 100,000 light-years. At this scale it would be 7,926 miles (12,756 km), equivalent to the diameter of the Earth.

The distance across the Orion nebula is 150 light-years. At this scale it is 12 miles (19 km), equivalent to the size of the city of Rome.

The diameter of the Moon is 2,160 miles (3,476 km). At this scale it is 0.000046 mm, equivalent to the size of a chromosome.

The scale of the universe is a concept that is hard to grasp. Its enormity can only really be understood by relating its scale to everyday objects. If the Galaxy is shrunk to a size equal to the diameter of the Earth, distances become easier to comprehend. For instance, the Earth would then only be as large as a microscopic virus, while the Sun would be the equivalent of a protozoan, a simple, single-celled animal. The edges of the planetary solar system would be a mere 3 inches (8 cm) from the Earth, and past this our nearest star, Proxima Centauri, would be just over ⅓ mile (0.5 km) away.

Beyond the limits of the Milky Way, our galactic neighbor M31, the galaxy in Andromeda, would be a disk slightly larger in diameter than the Earth and some 175,000 miles (281,000 km) away, or approximately three-quarters of the distance between the Earth and the Moon at full scale.

At the reduced scale, the whole of the Local Group of galaxies would be a disk nearly 560,000 miles (900,000 km) across, covering an area of space larger than that of the entire full-scale Earth-Moon system.

See also

OBSERVATORY EARTH
▶ Earth: astronomy's baseline 34/35

▶ Measuring the system 36/37

▶ Light-years 38/39

▶ Deep space 40/41

THE PLANETS
▶ Far out 98/99

SUN AND STARS
Stars in motion 110/111

NEBULAE AND GALAXIES
▶ Our galaxy 144/145

▶ Galactic neighbors 146/147

▶ Clusters and superclusters 148/149

HOW THE UNIVERSE WORKS
▶ In the beginning 178/179

The Planets

From the time when they were known only as points of light in the sky that moved across the fixed stars, the planets have fascinated human observers. We now know that these wandering stars, along with others too faint for the unaided eye to see, are a collection of unique worlds, some like our Earth, others vastly different. Looking at the planets from Earth, even with powerful telescopes, merely tantalized astronomers, so as soon as technology allowed, spacecraft were sent out. These probes have sent back a wealth of data, sometimes confirming, often exploding, previously held theories.

In addition to planets, however, there are also many smaller bodies orbiting the Sun, from specks of dust to chunks of rock and iron 600 miles (1,000 km) across. And then there are comets, occasional specters in the night sky, some of them thought to come from the remotest regions of the solar system, nearly halfway to the next star.

*Left (**clockwise from top**): space probe; Jupiter; magnetic Earth; spin and stretch; lunar features.*
***This page (top**): Venus revealed; **(left)** Saturn afloat; **(right)** life on Earth.*

Wandering stars

The idea that everything in the sky revolves around the Earth was confounded by tracking the planets.

Ancient peoples observing the sky over the course of months and years must have wondered why it was that not only the Sun and the Moon changed their position against the background of the apparently motionless stars. Five other starlike bodies moved around the heavens as well. From the earliest times, these "wandering stars," which we now know to be the five brightest planets (Mercury, Venus, Mars, Jupiter, and Saturn), were recognized as being different.

Philosophers tried to fit these movements into the prevailing view that the Earth was at the center of the universe and that everything in the heavens revolved around it. But this "geocentric" view gave Greek astronomers problems.

The Greeks believed in the perfection of the natural world, and this philosophy required the celestial bodies to travel in perfect circles. But these starlike objects do not move smoothly and regularly. The more astronomers watched the skies, the more irregularities came to light. The planets generally move in one direction, but occasionally appear to go backward (called retrograde motion), and their paths can look like a series of loops.

To preserve the idea of the perfection of nature, the Greeks assumed that the planets moved in smaller circles, or epicycles, as they moved in the larger circle that took them around the Earth. But the whole system of cycles and epicycles became immensely complex.

Eventually, it was deduced that the Earth and the planets move around the Sun in elliptical orbits. Since Venus and Mercury are seen only near the Sun, it is logical that they orbit closer to the Sun than Earth. The other planets are seen both close to and far from the Sun because they orbit farther from the Sun than Earth.

One difference between planets and stars is that only stars twinkle. This is because planets are close enough to show disks. Enough light rays from their disks reach our eyes without being refracted and bent by Earth's turbulent atmosphere, while light from remote, pointlike stars is dispersed, making the stars twinkle.

The plane of the Earth's orbit around the Sun is known as the plane of the ecliptic. The ecliptic can be projected onto the celestial sphere – the imaginary sphere (**bottom**) surrounding the Earth on which the heavenly bodies seem to appear – and marks the apparent yearly path of the Sun through the stars. The ecliptic is at an angle to the celestial sphere because the Earth's axis is tilted at an angle of almost 23½° to the plane of its orbit.

The rest of the planets also appear close to this line in a narrow belt known as the zodiac, as plotting the positions of Mars and Venus shows (**below**). The path of Mars is plotted in numbers and that of Venus in letters, each number or letter being the planet's position against the background of stars in successive months.

The 12 constellations that lie along the belt are known as the signs of the zodiac. The ecliptic was first divided into the 12 signs by the Babylonians as long ago as 3000 B.C.

Over a number of months, the Earth will catch up with and pass a planet farther from the Sun, such as Mars. This makes the track of the planet against the stars appear to move backward and loop *(below).*

THE CHURCH AND GALILEO

Galileo Galilei is renowned for his conflict with the Catholic Church over whether the Sun went around the Earth or the Earth went around the Sun. In 1632, when Galileo published his *Dialogue*, setting out the heliocentric (Sun-centered) theory, the idea had already been suggested by Aristarchus in the 3rd century B.C., by Copernicus in 1543, and proved by Kepler in 1609. The Church, however, had adopted the Earth-centered system expounded by 2nd-century astronomer Ptolemy.

The Inquisition made Galileo retract his views in 1633. But later his opinions, and those of other astronomers, became generally accepted.

Italian scientist Galileo *(1564–1642) was one of the founders of modern astronomy and one of the first to turn a telescope on the stars.*

An old plan of the heavens (right) shows, incorrectly, the Earth at the center of the universe.

The "superior" planets – the ones farther out than Earth from the Sun – travel around the Sun more slowly than the Earth. Because of this, the Earth catches up and overtakes them as it orbits, making them appear to travel backward – in an east-west, or retrograde, direction – before they revert to their regular west-east progression.

The closest superior planet is Mars *(right),* here caught making spectacular retrograde motions.

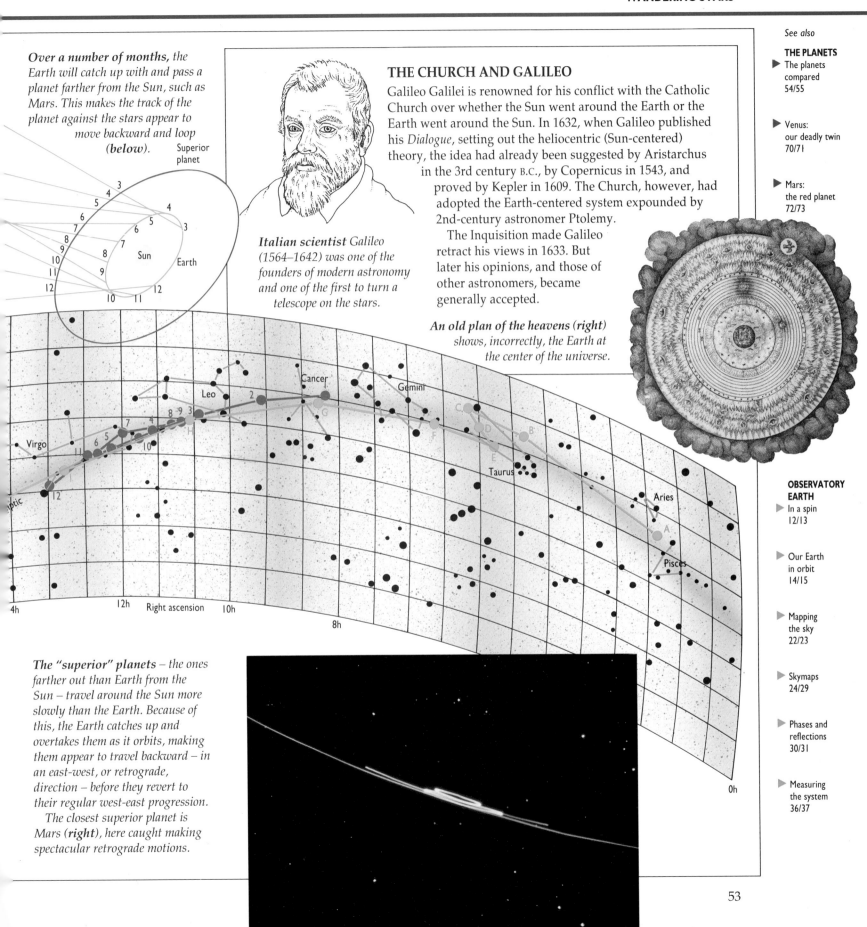

53

The planets compared

Despite differences in size and composition, the nine planets of the solar system have much in common.

Although the paths of the planets as they move across the sky look complicated when seen from Earth, if they were observed from the Sun, their individual paths through space would appear simple. This heliocentric, or Sun-centered, view shows the order and structure of the solar system and makes it obvious that it is not just a collection of random objects gathered together in space.

Each planet traces an elliptical – in most cases, an almost circular – orbit. All the planets orbit the Sun in the same direction, and all lie in roughly the same plane (the ecliptic). For instance, Venus's orbit is inclined at 3.4 degrees to the plane of the ecliptic and Mercury's at 7 degrees. The orbit of the outermost planet, Pluto, is inclined at 17 degrees, but even this is not much of a deviation. Furthermore, the planets are all almost spherical and seem to be roughly the same age and have the same origin. They all appear to have formed at about the same time as the Sun, some 4.5 billion years ago.

The planets are essentially of two sorts. The four closest to the Sun – Mercury, Venus, Earth, and Mars – are small, rocky, terrestrial, or Earthlike, planets with thin or negligible atmospheres. Heat from the young Sun blasted off most or all of any original gases, and the atmospheres now found on Venus, Earth, and Mars seeped from their cores later.

Beyond Mars is the asteroid belt, possibly the debris of a planet that never formed. Then come Jupiter, Saturn, Uranus, and Neptune, the four gas giants, so named because around their small, solid, rocky cores they have retained their huge, original atmospheres of light gases. Much of the gas around each giant's core is compressed, liquefied, or even made semi-solid by the phenomenal pressure of a deep atmosphere. They all have rings and complex moon systems.

Beyond Neptune is the oddball, Pluto, which is made mainly of ice and frozen methane and ammonia. It has only one moon, Charon, which is fully half its size.

CHARTING THE DIFFERENCES

	Mass (Earth = 1)	Mean density (g/cm³)	Diameter (Earth = 1)	Gravity on equator (Earth = 1)	Period of rotation
Mercury	0.055	5.43	0.38	0.28	58.65 Earth days
Venus	0.82	5.25	0.95	0.88	243.01 Earth days
Earth	1	5.52	1	1	23.93 Earth hours
Mars	0.11	3.95	0.53	0.38	24.62 Earth hours
Jupiter	317.94	1.33	11.2	2.34	9.84 Earth hours
Saturn	95.18	0.69	9.45	0.93	10.23 Earth hours
Uranus	14.53	1.29	4.01	0.79	17.9 Earth hours
Neptune	17.14	1.64	3.88	1.12	19.2 Earth hours
Pluto	0.0022	2.03	0.18	0.04	6.39 Earth hours

On Mars, which has only about one-ninth the mass of Earth, the force due to gravity is one-third of that on our planet. One pear on Earth has a weight equivalent to that of nearly three pears on Mars (above left). While the mass of a pear is the same whether it is on Mars or Earth, its weight depends on the force of gravity acting on its mass.

Gravity depends not only on the mass of a body, but also on how close an object is to the center of the body's mass. Because Mars is smaller than the Earth, an object on Mars's surface is closer to the center of the planet and therefore to the center of mass. This means that the force of gravity is felt more strongly.

Uranus

If Jupiter were as big as an orange, then Earth would be the size of a pea (left). Jupiter's diameter is about 11 times that of the Earth, so the planet's volume is about 1,300 times Earth's. But because Jupiter is not very dense, it is only some 320 times the mass of the Earth. Earth is composed mainly of rock and metal, while Jupiter is made mainly of gas and liquefied gas.

Gas giant Saturn has such a low density that a piece of it would float in water, whereas a lump of Earth matter would sink (right).

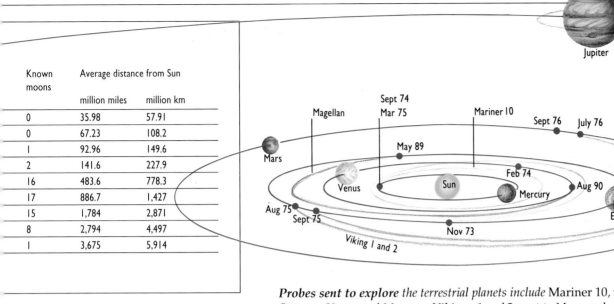

Known moons	Average distance from Sun	
	million miles	million km
0	35.98	57.91
0	67.23	108.2
1	92.96	149.6
2	141.6	227.9
16	483.6	778.3
17	886.7	1,427
15	1,784	2,871
8	2,794	4,497
1	3,675	5,914

Probes sent to explore the terrestrial planets include Mariner 10, *which flew past Venus and Mercury, Vikings 1 and 2,* sent to Mars, and Magellan, *sent to orbit Venus. Vikings 1 and 2 were launched from Earth in August and September 1975 and arrived at Mars a year later. Each Viking was made up of a lander and an orbiter.*

Most of the satellite probes launched to examine the solar system carry TV cameras to send back views of the planets. Some also carry ultraviolet or infrared cameras, or radar equipment to see through clouds. Instruments can include magnetic field detectors and receivers to pick up radio signals.

Once probes have left Earth, they are effectively in free-fall – powered by their momentum and the force of gravity. Paths are planned for multiplanet visits, and probes have used the pull of gravity from each planet in turn to pick up speed and propel them on to the next planet. Voyager 2, for instance, managed to call at Jupiter, Saturn, Uranus, and Neptune before leaving our solar system.

Planet Earth

Aliens in the neighborhood of the Sun would probably home in on Earth.

If you were an alien coming from outer space on a mission to boldly go where none of you had been before, you might well be attracted to our solar system to seek out a new world or civilization. You would most likely detect radio signals that have been coming from our solar system for around 100 years and have spread nearly 560 trillion miles (900 trillion km) out into space.

The complex patterns in these signals would indicate that they are not from a natural source and that they are the product of an "intelligent" civilization. Since the 1930s, TV pictures have also been traveling out into space and, if you had the technical knowhow and expertise to travel through interstellar space, you would probably be able to view them to see what we look like.

PLANET PROFILE: EARTH

		miles	km
Distance from Sun	max.	94.4 million	152 million
	min.	91.3 million	147 million
	ave.	93 million	150 million
Equatorial diameter		7,926	12,756
Polar diameter		7,900	12,714
Escape velocity		7.02/sec	11.3/sec
Mean orbital velocity		18.5/sec	29.8/sec
Surface temperature	ave.	59°F	15°C
Atmospheric pressure (at surface)		14.7 lb/sq in	760 mmHg
Length of year		365.26 days	
Rotation period		23.93 hours	
Mass		5.976×10^{21} tons	
Mean density		5.52 g/cm³	
Orbital eccentricity		0.017	
Angle of equator to orbit		23.45°	
Albedo (reflectivity)		0.39	
Gravity at equator		32 ft/sec⁻²	9.78 m/sec⁻²

We know more about the interior of the Earth than we do about that of any other body in the universe. The crust on which we live is just 4–25 miles (6–40 km) thick, a fraction of the diameter of the planet and thinner in proportion than an eggshell is to an egg.

From what we know about the Earth it has been possible to surmise much about the interiors of the other planets in the solar system, especially the rocky, Earthlike, inner planets.

— Crust
— Upper mantle
— Mantle
— Inner core
— Outer core

Earth rising over the Moon might be the first close-up view of our planet to alien prospectors lured to our solar system by the radio noise we broadcast into space. Approaching from behind the Moon, aliens would observe the constantly changing cloud patterns of the Earth's weather systems. Underneath these, the landmasses and the oceans show clearly.

At the poles of this predominantly blue world are the white patches typical of frozen material, in Earth's case, water.

Even before you entered our solar system, you would be able to work out how much heat the Sun gave out and would be looking for a planet roughly in the region of the Earth's orbit. Only a planet at about this distance from the Sun would have water in its liquid state on its surface – and liquid water is thought to be vital to life, wherever in the universe it might originate. Furthermore, the orbit of the Earth around the Sun is almost circular, so the conditions on the planet's surface are relatively stable.

As you passed the orbit of Pluto, you would be able to see that the Earth is a double planet, the other planet (our moon) being about one-eightieth of the size of the main planet it orbits.

Analysis of the sunlight reflected from the Earth would tell you the composition of its atmosphere. The most common gas in the air is nitrogen (78 percent). But from the high proportion of oxygen (21 percent) and low proportion of carbon dioxide (about 450 parts per million), you would be able to tell that there were carbon-based life forms on Earth using oxygen to power their energy-production systems. Key mechanisms in maintaining that balance in the air are plants, which photosynthesize, absorbing carbon dioxide and producing oxygen, and crustaceans, which use carbon dioxide in building their shells.

After the most cursory examination of the planet you would see that some 71 percent of its surface is covered with water. And above the surface are huge swirling clouds, which are made of water vapor. So with the liquid water on the surface and an atmosphere that pointed to the fact that life of some sort was busy going about the process of living, you would certainly have no doubt at all that this planet was the home of those radio and TV signals.

You would then have to decide whether or not you wanted to meet the beings who broadcast all those signals out into space. That might not be such an easy decision to make.

EARTH'S MAGNETIC FIELD

One of the features of Earth that aliens surveying the planet from afar would notice would be its magnetic field. On Earth, we use it to navigate. A magnetic compass (**above**) points to the Earth's magnetic north pole – a region of the planet's surface where the lines of magnetic force point directly downward.

The Earth itself acts as a large magnet, with a north and south pole. Most of its magnetism is generated by electric currents in its core, though around 10 percent is made by currents in the ionosphere in the upper atmosphere.

The magnetic poles roughly coincide with the geographic poles, the end points of the Earth's axis of rotation. They are only about 1,000 miles (1,600 km) apart. The magnetic north pole is currently in Viscount Melville Sound in northern Canada. The magnetic south pole is off the coast of Antarctica in Australian territory. The magnetic poles change position slowly, but never move far from the geographic poles, though their position relative to the continents has changed over geological time due to continental drift.

Layer of life

Protected by nothing more than the air above, life thrives in a thin, habitable skin around the Earth.

Coming in past the orbit of the Moon, visitors to Earth would begin to see patches of vegetation. But they would have to be in orbit at around 300 miles (480 km) – the space shuttle's maximum altitude – before they could see signs of human life.

First, they would see the lights of major cities at night. Then, as they came lower, the wakes of huge oil tankers plying the oceans would come into view. Lower still, geometric, cultivated fields, rail tracks, roads, and urban sprawl would be distinguishable.

All life on Earth, not just human life, is confined to a narrow layer less than 12 miles (20 km) deep. Living organisms exist in the atmosphere up to around 5 miles (8 km), on the Earth's surface itself, in the top layer of the soil, in caves, and in the oceans down to about 7 miles (11 km) below sea level. Within this narrow shell around the Earth, life is extraordinarily abundant.

A crucial factor in the development of life is the Earth's atmosphere, which is held on the planet's surface by gravity. It contains 78.09 percent nitrogen, 20.95 percent oxygen, 0.93 percent argon, 0.03 percent carbon dioxide, small amounts of water vapor, and traces of methane and other gases.

The primordial atmosphere, which came from gases belched out by volcanoes and undersea vents, contained 100 times the amount of carbon dioxide and little free oxygen. Much of the carbon dioxide has been extracted from the air by weathering of rocks to make limestone. Photosynthesis by green plants has also absorbed carbon dioxide and, along with the breakdown of water vapor by sunlight, released oxygen into the air.

All the varied habitats and the organisms that live in them depend on the existence of an atmosphere. It provides the oxygen in the air that we breathe and acts as a blanket, keeping the surface warm enough for liquid water to exist.

Visible light from the Sun passes through the atmosphere and warms the Earth's surface. It is then re-radiated from the surface at longer, infrared, wavelengths, which do not pass through the atmosphere so easily. The principal so-called greenhouse gases – carbon dioxide, methane, and water vapor – absorb this energy and re-emit it back down to the surface, effectively trapping it and keeping the planet 54°F (30°C) warmer than it would otherwise be. This effect is largely beneficial, maintaining the surface of the Earth at temperatures where water is liquid and life is possible.

The atmosphere acts as an effective barrier to most solid matter, and all but the biggest meteors burn up in the upper atmosphere. It also keeps out dangerous radiation, such as X-rays, that are harmful to life.

Key players in this latter role are four electrically charged layers in the ionosphere – the D, E, F¹, and F² layers – at 50, 70, 125, and 250 miles (80, 110, 200, and 400 km) respectively. They are formed by the breakdown of gas atoms into free electrons and positively charged ions by the energy in sunlight.

Aurorae – visible in the night sky in polar regions – are caused by charged solar particles colliding with molecules in the Earth's atmosphere. These collisions cause the molecules to reach an excited state and emit photons of energy, which are the colors that can be seen in the Northern and Southern lights.

Ozone occurs in a diffuse layer in the stratosphere between 12 and 30 miles (20 and 50 km) above the Earth's surface. It plays an important role in absorbing most of the ultraviolet radiation that reaches us from the Sun. If this radiation were to reach ground level, it would prove fatal to most forms of life.

Meteor

125 miles (200 km)

Earth's weather almost all takes place in the troposphere.

250 miles (400 km)

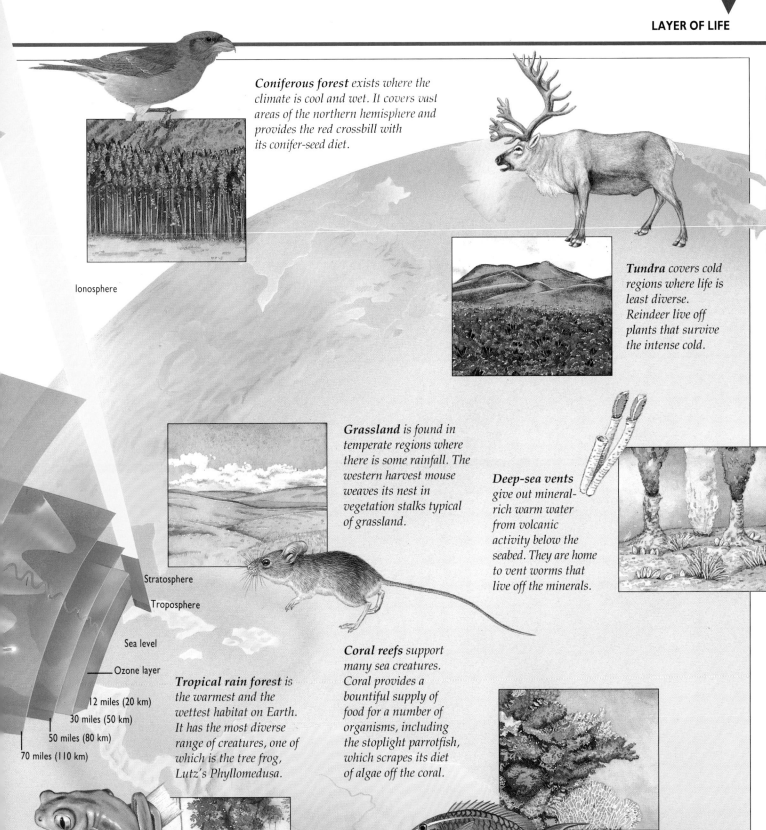

Coniferous forest *exists where the climate is cool and wet. It covers vast areas of the northern hemisphere and provides the red crossbill with its conifer-seed diet.*

Ionosphere

Tundra *covers cold regions where life is least diverse. Reindeer live off plants that survive the intense cold.*

Grassland *is found in temperate regions where there is some rainfall. The western harvest mouse weaves its nest in vegetation stalks typical of grassland.*

Deep-sea vents *give out mineral-rich warm water from volcanic activity below the seabed. They are home to vent worms that live off the minerals.*

Stratosphere

Troposphere

Sea level

Ozone layer

12 miles (20 km)

30 miles (50 km)

50 miles (80 km)

70 miles (110 km)

Coral reefs *support many sea creatures. Coral provides a bountiful supply of food for a number of organisms, including the stoplight parrotfish, which scrapes its diet of algae off the coral.*

Tropical rain forest *is the warmest and the wettest habitat on Earth. It has the most diverse range of creatures, one of which is the tree frog, Lutz's Phyllomedusa.*

The inside story

The apparent solidity of the ground we walk on might make us think that the planet under our feet is stable and inert. But much is going on below the surface.

Earth's rocky surface is just a thin crust floating on molten liquid rock, rather like the skin on a rice pudding. The crust is divided into nine main regions, called tectonic plates (plus a number of smaller ones), on which sit the continents and smaller masses of land. Over time the plates, and the continents they carry, have drifted over the Earth's surface.

The theory that the continents move was first proposed in about 1800 when it was noted that the eastern bulge of South America "fitted" the shape of the bight of Africa. Identical fossils were found in Europe and North America, indicating that the two continents had at one time been joined. In the 1900s, continental collision was used to explain the formation of mountain ranges.

But the first comprehensive theory of continental drift was suggested by German meteorologist Alfred Wegener in 1912. He proposed that throughout most of geological time, there had been only one continent, called Pangaea, surrounded by a single ocean, Panthalassa. Pangaea, he maintained, broke up in the Jurassic period, 210 to 140 million years ago. In 1937, South African geologist Alexander du Toit modified the theory, suggesting that there may have been two protocontinents: Laurasia,

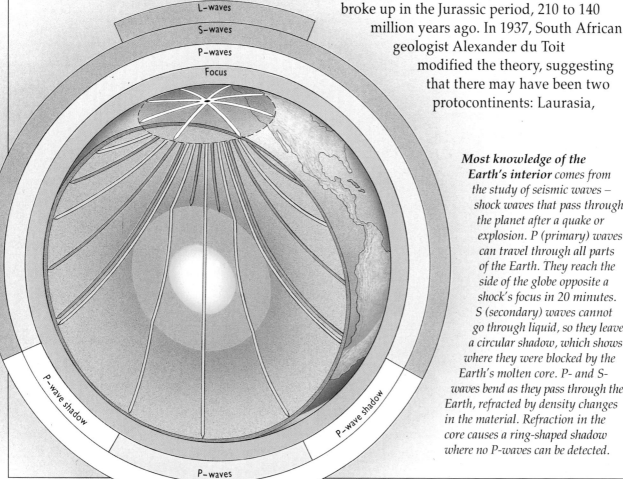

Most knowledge of the Earth's interior comes from the study of seismic waves – shock waves that pass through the planet after a quake or explosion. P (primary) waves can travel through all parts of the Earth. They reach the side of the globe opposite a shock's focus in 20 minutes. S (secondary) waves cannot go through liquid, so they leave a circular shadow, which shows where they were blocked by the Earth's molten core. P- and S- waves bend as they pass through the Earth, refracted by density changes in the material. Refraction in the core causes a ring-shaped shadow where no P-waves can be detected.

When the Earth reminds us that its crust is just a thin, unstable layer, the results can be devastating. Where plates collide or rub against each other, they sometimes stick. As the stored energy of the movement builds up, the stresses at the boundary between the plates increase. When the plates unstick and movement finally comes, it can cause a cataclysmic earthquake. The aftermath of a shock at Gibellina, Sicily (above), shows how extensive destruction can be.

Most damage is caused by seismic waves called Love waves, or Rayleigh waves (L-waves). They travel in the Earth's surface near a focus (left), causing the ground to shake from side to side or move up and down.

L-waves

S-waves

P-waves

Focus

P-wave shadow

P-wave shadow

P-waves

S-wave shadow

which broke up to form the northern continents – North America, Europe, and Asia except for the Indian subcontinent; and Gondwana, which included South America, Africa, India, Australia, and Antarctica. Laurasia and Gondwana were separated by the Tethys Ocean, which has since largely closed, leaving behind the Mediterranean Sea. Continental drift continues: North America is moving away from Europe at a rate of ¾ inch (2 cm) a year.

The continental crust is mostly made of granite. It supports the continents and is on average 20 miles (33 km) thick, although it may reach over 37 miles (60 km) under mountain ranges. The oceanic crust is much thinner, at between 4 and 10 miles (6 and 16 km) thick, and is made largely of basalt and gabbro, both types of heat-formed, or igneous, rock. Oceanic crust is newer than continental crust since it is constantly being made along ridges in the middle of the oceans.

Below the crust is the mantle, which is thought to be made of peridotite, a rock composed mainly of magnesium and iron silicates. Convection currents in the molten mantle material are believed to cause new material to erupt at the ocean ridges. The new sea floor spreading out from the ridges then drives the process of continental drift; as the oceans widen, the Earth's plates are forced apart. This, in a nutshell, is the theory of plate tectonics (broadly accepted since the 1960s), which has provided a convincing explanation for the causes of continental drift.

Under the mantle, about 1,800 miles (2,900 km) below the Earth's surface, is the core. The inner core is thought to be made of solid iron some 1,490 miles (2,400 km) across. Around it is an outer core, some 1,240 miles (2,000 km) thick, made of molten iron and nickel, probably mixed with sulfur and oxygen. Electric currents in this outer core are believed to cause the Earth's magnetic field.

When magma, or molten rock, wells up beneath a continent from hot regions of the mantle (**below right**) *it can force the crust to separate. Where magma wells up through the floor of an ocean* (**below left**) *it creates new ocean crust and causes sea-floor spreading. At regions called subduction zones, crust on one plate* is forced under crust on another and a deep ocean trench forms. Where a plate divides under a continental plate there are land-based volcanoes, mountains, and earthquakes.

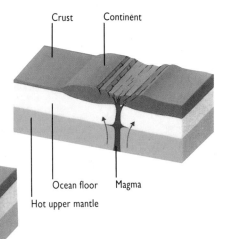

Crust   Continent

Ocean trench

Cold upper mantle

Ocean floor   Magma

Hot upper mantle

Magma

Companion planet

Gaze skyward one night and you might see one of the solar system's most beautiful sights – the Moon, Earth's close astronomical partner.

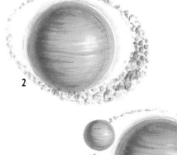

1

2

3

None of the other inner, terrestrial planets has a moon like ours. Mercury and Venus have no moons at all, and Mars is orbited only by two tiny chunks of rock, each just one ten-millionth the size of our moon.

Earth and Moon, held together by their mutual gravitational attraction, orbit around a common center of mass. Because the Moon rotates on its axis in exactly the time it takes to orbit the Earth, it always presents the same side to us. But since the Moon's orbit is slightly elliptical, it speeds up and slows down during its orbit and seems to wobble from side to side, revealing an extra 16 degrees of its surface.

The Moon does not emit light – what we see is sunlight reflected from the Moon's surface as the Moon goes through its familiar phases. At first quarter, it is in the sky at dusk and sets about halfway through the night, while at last quarter it does not rise until midnight. The old or new Moon can be seen only low in the sky near dawn or dusk. The Moon is in the sky all night for a few days a month when near full phase on the opposite side of the Earth from the Sun. The Moon is best observed near its quarter phases, when the long shadows cast by the Sun on the Moon's surface make features stand out. The Moon can also be observed near its new phase, when lit by earthshine.

The Moon is too small to retain much atmosphere, and the atmospheric pressure at its surface is only about 10^{-14} of the Earth's – so low as to be effectively nonexistent.

Core

Mantle

Upper crust
Lower crust

How the Moon formed is not known, but the most widely accepted hypothesis is the "splash" theory. According to this, Earth was hit early in its history by a huge object (1) and material splashed from Earth's molten surface to make a ring (2). The debris in the ring then came together to create the Moon (3).

Alternatively, it may have formed from the accretion of debris and dust around the Earth while Earth itself was coalescing.

Another theory holds that the Moon originated in another part of the solar system and was captured by the Earth's gravity as it flew by. But the mechanics of such a capture make this an unlikely scenario.

PLANET PROFILE: THE MOON

		miles	km
Distance from Earth	max.	252,710	406,697
	min.	221,463	356,410
	ave.	238,607	384,000
Diameter		2,160	3,476
Escape velocity		1.48/sec	2.38/sec
Mean orbital velocity		2,287/hour	3,680/hour

	day	hour	min	sec
Time to spin on axis	27	7	43	11.5
Time to orbit Earth	27	7	43	11.5
Time to go through phases	29	12	44	2.8

Surface temperature	°F	−292° to +248°
	°C	−180° to +120°

Mass	$\frac{1}{81}$ of Earth's
Density	0.6 of Earth's
Angle of orbit to ecliptic	5.15°
Angle of equator to ecliptic	1.53°
Apparent diameter	0.518° (average)
Magnitude of full Moon	−12.7
Albedo (reflectivity)	0.073
Gravity at equator	0.166 of Earth's

A drop falling into water makes a crater shape for an instant before the energy is dissipated as ripples. Moon rock melted by the impact of a chunk of debris quickly solidifies, freezing the crater shape in solid rock.

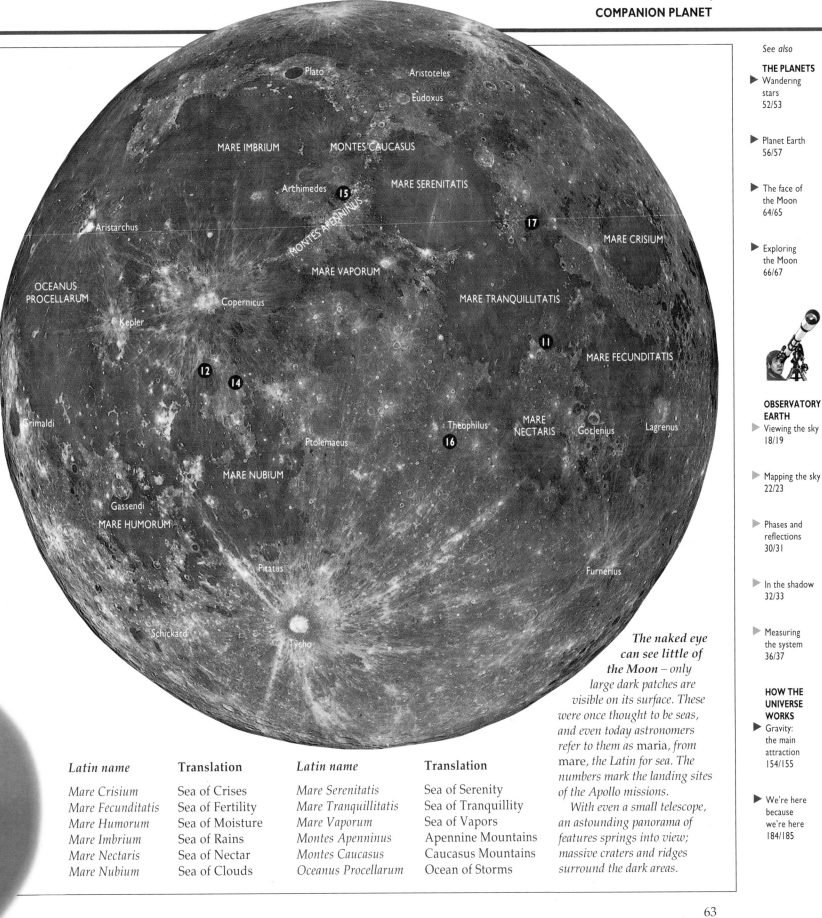

Plato

Aristoteles

Eudoxus

MARE IMBRIUM   MONTES CAUCASUS

Arthimedes   **15**   MARE SERENITATIS

Aristarchus

MONTES APENNINUS

17

MARE CRISIUM

MARE VAPORUM

OCEANUS
PROCELLARUM

Copernicus   MARE TRANQUILLITATIS

Kepler

11

MARE FECUNDITATIS

12

14

Grimaldi

Theophilus   MARE
NECTARIS   Goclenius   Lagrenus

Ptolemaeus   **16**

MARE NUBIUM

Gassendi

MARE HUMORUM

Pitatus   Furnerius

Schickard

Tycho

Latin name	Translation	Latin name	Translation
Mare Crisium	Sea of Crises	*Mare Serenitatis*	Sea of Serenity
Mare Fecunditatis	Sea of Fertility	*Mare Tranquillitatis*	Sea of Tranquillity
Mare Humorum	Sea of Moisture	*Mare Vaporum*	Sea of Vapors
Mare Imbrium	Sea of Rains	*Montes Apenninus*	Apennine Mountains
Mare Nectaris	Sea of Nectar	*Montes Caucasus*	Caucasus Mountains
Mare Nubium	Sea of Clouds	*Oceanus Procellarum*	Ocean of Storms

*The naked eye
can see little of
the Moon* – only
*large dark patches are
visible on its surface. These
were once thought to be seas,
and even today astronomers
refer to them as* maria, *from*
mare, *the Latin for sea. The
numbers mark the landing sites
of the Apollo missions.*

*With even a small telescope,
an astounding panorama of
features springs into view;
massive craters and ridges
surround the dark areas.*

See also

THE PLANETS
▶ Wandering
stars
52/53

▶ Planet Earth
56/57

▶ The face of
the Moon
64/65

▶ Exploring
the Moon
66/67

**OBSERVATORY
EARTH**
▶ Viewing the sky
18/19

▶ Mapping the sky
22/23

▶ Phases and
reflections
30/31

▶ In the shadow
32/33

▶ Measuring
the system
36/37

**HOW THE
UNIVERSE
WORKS**
▶ Gravity:
the main
attraction
154/155

▶ We're here
because
we're here
184/185

The face of the Moon

Rocky surface features on the Moon have not always been the same – the way they have changed in the past 4.5 million years is to be seen on the Moon of today.

4 billion years ago

Only after men landed on the Moon and gathered rocks from the surface was it possible to work out the Moon's history with any degree of accuracy. Evidence from rock samples has been combined with analysis of the Moon's landscape to reveal the violent, convulsive secrets behind its appearance.

The Moon's surface is covered by a layer made of pulverized rock called regolith. Lunar rock is similar to rock on Earth. The main difference is that lunar rock is richer in metals such as titanium that have a high boiling point. Substances with a relatively low boiling point, such as water, are absent. They evaporated into space long ago.

The paler areas are the highlands. They are more heavily cratered and are thus thought to be older than the darker "seas" they surround because they have had more time to be hit by meteors. The highlands, which can rise to over 20,000 ft (6,000 m) above the seas, were created when the Moon first formed, 4.5 billion years ago. Many of them are breccias – fragments of different types of rock welded together by high-velocity impacts. The highlands have a higher aluminum and calcium content than the seas, which gives them their lighter color.

Beginning some 3.6 billion years ago, lowland areas were covered by molten lava, which welled up following the melting of rock layers 125 miles (200 km) below the Moon's surface. This upwelling covered the previously heavily cratered lowlands to create the darker, relatively uncratered seas. These are made of basalt, similar to dark volcanic rocks on Earth.

The features of the Moon are, in fact, a mix of impact craters and volcanic features such as volcanic craters, lava flows, rills, and fault lines. One dramatic fault line is the Rupes Recta, a wall-like cliff 820 ft (250 m) high and 80 miles (129 km) long in the east of the Mare Nubium, visible with a small telescope.

About 4 billion years ago, after the Moon had assembled from debris and its surface had cooled down a little, it looked similar to the reconstruction shown above. Impacts of chunks of matter, some tiny, some up to 155 miles (250 km) wide, had formed craters ranging from the size of a pinhead to hundreds of miles across. But some of these craters did not last long, as the Moon entered a new era, one in which volcanism played a major part.

TAKE A LOOK AT THE FAR SIDE

Nearly half of the Moon's surface remained hidden from human view until October 1959, when Soviet space probe *Luna 3* made a successful orbit of the Moon and sent back the first pictures of the far side.

Between August 1965 and August 1966, American lunar orbiters made a complete survey of the Moon – including the far side – to prepare for the manned missions. The photograph (**left**) was taken during the *Apollo 16* mission in 1972.

Despite much speculation about what it might look like, the far side revealed few surprises. Like the near side, the far side of the Moon is covered with craters, highlands, and seas. But the seas are smaller and the highlands more extensive. Nor are there huge oceans to match the Oceanus Procellarium or the Mare Imbrium.

Terraced crater

The Moon's craters range in size from microscopic pits to huge depressions up to 155 miles (250 km) across. Most craters are formed by the impact of bodies on the Moon's surface. The energy of the incoming body is converted to heat that melts the surface for an instant before it solidifies in a crater shape. Terraced craters, some of the biggest on the Moon, are formed by the impact of large bodies. Their terraced walls and central peaks are caused by material rebounding from the edge of the craters after impact. The central peak is a frozen splashback of rock. Terraces only occur on impact craters larger than 12½ miles (20 km) across. Impact craters below this size are bowl-shaped.

Not all craters result from impacts: concentric craters are volcanic sites where lava has been emitted from below. Ghost craters are a hybrid: they are impact craters that have been swamped by lava that spread over the Moon's surface until only a trace of their original outline protrudes through the lunar sea.

Ray craters are relatively recent impact craters. The rays are material ejected at the time of impact that is lighter in color than the underlying, older mare matter. Some of the most interesting features on the Moon are rills, which look like dried up riverbeds up to 300 miles (500 km) long. Some areas are free of them; others are crisscrossed. Rills seem to be the channels of empty lava tubes whose roofs have collapsed. Similar features are found on Earth.

Concentric crater

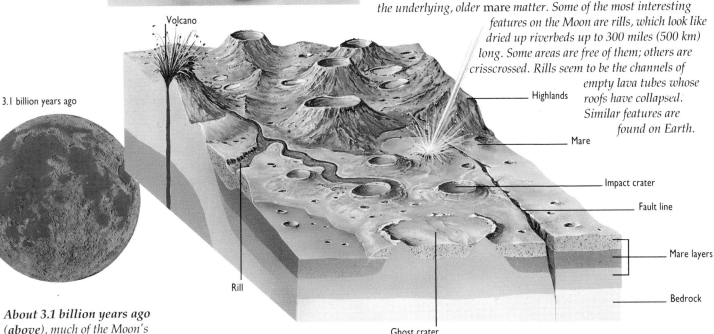

Volcano

Highlands

Mare

Impact crater

Fault line

Mare layers

Bedrock

Ghost crater

Rill

3.1 billion years ago

About 3.1 billion years ago (above), much of the Moon's surface had been "smoothed out" by volcanic activity which spewed molten lava into the low-lying regions, filling many of the large craters to make the seas.

The Moon today (right) looks much the same as it did after the maria, or seas, appeared. Any changes are due to impact by stray chunks of material which formed craters on the otherwise featureless maria and also on already cratered regions.

Present day

Ghost crater

Ray crater

Exploring the Moon

For centuries, people have dreamed of traveling to the Moon. Yet only in 1969 was this ambition finally fulfilled.

The Apollo space program was announced by President John F. Kennedy in a speech to Congress in May 1961. Its aim was to put a man on the Moon "before the decade was out." Scientists from NASA achieved it within eight years.

German rocket pioneer Wernher von Braun was employed to build a huge three-stage launch vehicle – the *Saturn V* – and, on October 11, 1968, following a number of unmanned flights, the first three-man crew was blasted into Earth orbit on *Apollo 7*. Subsequent manned missions allowed checks to be made on the equipment and docking procedures in both Earth and lunar orbits.

Then, on July 21, 1969, during the *Apollo 11* mission, astronaut Neil Armstrong became the first man to set foot on the Moon.

Apollo 11's *aim* was simply to put a man on the Moon, and it only carried two experiments. One, a seismometer to measure moonquakes, was deployed by Edwin "Buzz" Aldrin (**below**). The other was a laser reflector used to measure the distance between Earth and Moon accurately. Later missions were more fully equipped. Eugene Cernan, who flew on Apollo 17, is shown testing the lunar rover (**bottom left**), which gave astronauts greater freedom to explore the Moon.

THE APOLLO MISSIONS

After liftoff (**1**), the third stage of the *Saturn V* rocket enters Earth orbit, having jettisoned its first and second stages. The third stage is then re-ignited (**2**), blasting the spacecraft out of Earth orbit, whereupon the linked command and service modules turn around (**3**) to dock with the lunar module (**4**). After $2\frac{1}{2}$ days, the modules are close

Edwin Aldrin (11)

Neil Armstrong (11)

Allan Bean (12)

Charles Conrad Jnr. (12)

Edgar Mitchell (14)

See also

THE PLANETS
▶ Planet Earth
56/57

▶ Companion
planet
62/63

▶ The face of
the Moon
64/65

**OBSERVATORY
EARTH**
▶ Phases and
reflections
30/31

▶ In the shadow
32/33

▶ Measuring
the system
36/37

**SUN AND
STARS**
▶ Center of
the system
108/109

**HOW THE
UNIVERSE
WORKS**
▶ Getting heavy
156/157

Lunar rovers were carried by the last three Apollo missions – 15, 16, and 17. With a top speed of 7½ mph (12 km/h), the astronauts could cover more than 20 miles (30 km) in one session outside the spacecraft using the rover. This extended range allowed a greater diversity of rock samples to be collected (**left**), including an interesting type of orange soil, which was eventually found to contain ancient, colored glassy particles.

With no weather to disturb them, the footprints and tire tracks left by the Apollo missions will remain on the face of the Moon for millions of years.

Interest in lunar exploration waned after the last of the Apollo missions in 1972, and only three more probes, all Soviet, have been to the Moon – the latest in 1974.

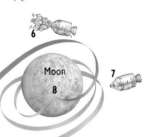

However, recent studies have been made regarding the viability of a new wave of lunar exploration. Proposals have included establishing a lunar base to mine the minerals, such as titanium, found in the Moon's rocks.

to the Moon (**5**), and the rockets are fired to enter a lunar orbit (**6**). Leaving the command module in orbit (**7**), the lunar module touches down on the Moon (**8**). Once finished there, the top half of the lunar module lifts off to rendezvous with the command module (**9**). The spacecraft then leaves the Moon's orbit for the return journey (**10**). On approaching Earth, the command module separates from the service module (**11**) and begins its descent (**12**). At 75 miles (120 km), the command module enters the Earth's atmosphere (**13**). Friction with the atmosphere heats up the module (**14**), which is protected by a heat shield. Finally, it splashes down in the ocean.

All 12 astronauts who have walked on the Moon are shown below along with their mission numbers.

Alan Shepard (14)

James Irwin (15)

David Scott (15)

Charles Duke Jnr. (16)

John Young (16)

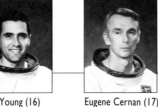
Eugene Cernan (17)

Harrison Schmitt (17)

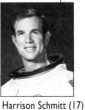

Mercury: fire and ice

Because of its closeness to the Sun, Mercury is a truly dead world,
alternately roasting and freezing as it slowly rotates on its axis.

Early civilizations thought that Mercury was two stars, since it can only be seen low in the sky at dawn or dusk. But the Greeks realized that it was a single planet and called it Hermes, which was their name for the god the Romans called Mercury.

Little bigger than our moon, Mercury is the smallest of the solar system's eight major planets. It looks like the Moon, too, with its cratered surface. Mercury has almost no atmosphere to protect its surface from the impact of meteors and to erode the resulting craters. Any atmosphere Mercury may once have had has been blasted away by the stream of particles coming from the nearby Sun or boiled off into space by the intense heat. What remains is helium at a pressure only a tiny fraction of that on Earth's surface.

With no atmosphere to shield or blanket it, the temperature reaches 806°F (430°C) during the day, hot enough to melt lead, and at night

Mantle
Crust
Core

The surface features of Mercury were seen in detail for the first time in images sent back by the American space probe Mariner 10, *launched in November 1973. This photograph (**below**), taken on March 29, 1974, reveals the planet's lifeless, crater-ravaged exterior.*

Mariner 10, now inactive, is still in orbit around the Sun. No further probes to Mercury are planned, and manned missions are unlikely because of the extreme conditions that would be encountered there.

PLANET PROFILE: MERCURY

		miles	km
	max.	43.3 million	69.7 million
	min.	28.5 million	45.9 million
Distance from Sun	ave.	36 million	57.9 million
Diameter		3,031	4,878
Escape velocity		2.7/sec	4.3/sec
Mean orbital velocity		29.8/sec	47.9/sec
Length of year		87.97 Earth days	
Rotation period		58.65 Earth days	
Surface temperature	°F	−292 to +806	
	°C	−180 to +430	
Atmospheric pressure	(at surface)	2×10^{-12} of Earth's	
Mass		0.055 of Earth's	
Orbital eccentricity		0.206	
Angle of equator to orbit		2°	
Albedo (reflectivity)		0.07	
Gravity at equator		0.28 of Earth's	

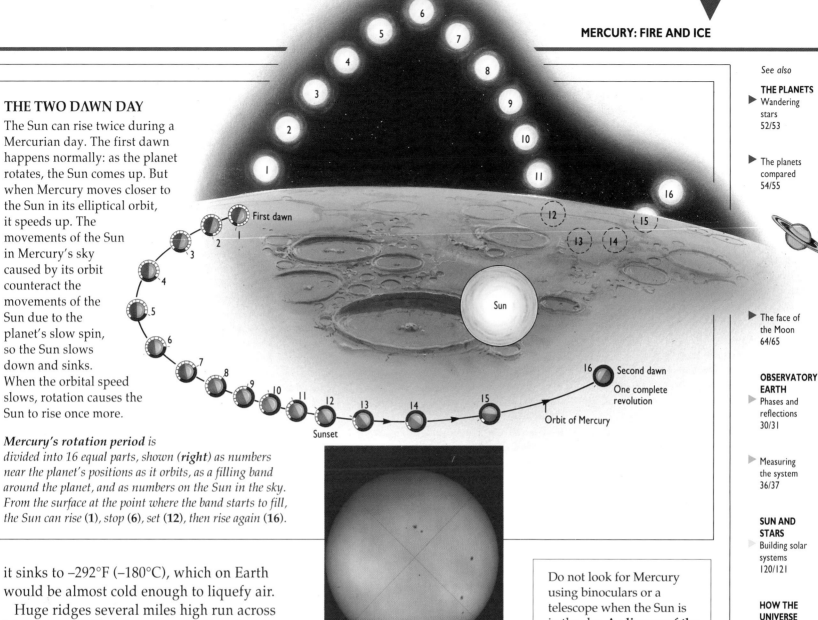

THE TWO DAWN DAY

The Sun can rise twice during a Mercurian day. The first dawn happens normally: as the planet rotates, the Sun comes up. But when Mercury moves closer to the Sun in its elliptical orbit, it speeds up. The movements of the Sun in Mercury's sky caused by its orbit counteract the movements of the Sun due to the planet's slow spin, so the Sun slows down and sinks. When the orbital speed slows, rotation causes the Sun to rise once more.

Mercury's rotation period is divided into 16 equal parts, shown (right) as numbers near the planet's positions as it orbits, as a filling band around the planet, and as numbers on the Sun in the sky. From the surface at the point where the band starts to fill, the Sun can rise (1), stop (6), set (12), then rise again (16).

First dawn

Sun

Second dawn
One complete revolution

Orbit of Mercury

Sunset

it sinks to –292°F (–180°C), which on Earth would be almost cold enough to liquefy air.

Huge ridges several miles high run across Mercury's surface. It is thought that they were created after the planet formed, when its huge nickel-iron core shrank by several miles and the surface wrinkled around it. Other surface buckles have been caused by shock waves from meteorite impacts. The heaviest bombardment of Mercury took place a few hundred million years after it was formed, and some of the craters are several hundred miles across. Volcanoes were active on the planet at that time, and some craters are flooded with lava. However, there are no active volcanoes on Mercury now, and the planet's surface has changed little in the last 3 to 4 billion years.

Mercury is dwarfed by the Sun, like all the planets. This can be seen most clearly when the tiny disk of the planet is visible against the raging furnace of the Sun's exterior. The planet is the small dot just below the point where the lines superimposed on the Sun intersect (**above**). Mercury can be seen in silhouette traveling across the Sun about 13 times a century.

The movement of a smaller body across the face of a larger one is called a transit. The next transits of

Do not look for Mercury using binoculars or a telescope when the Sun is in the sky. **A glimpse of the Sun will cause eye damage and possibly blindness.**

Mercury in front of the Sun, as seen from the Earth, occur on November 15, 1999, and May 5, 2003.

Although it has the fastest orbital velocity of all the planets, at an average of about 107,000 mph (172,000 km/h), it can take up to nine hours for Mercury to complete a transit. But then, the Sun's diameter is nearly 870,000 miles (1.4 million km).

Venus: our deadly twin

Once thought to be a steamy, jungle world, Venus is now known to be a choking, high-pressure inferno.

In size, shape, and mass, Venus is roughly the Earth's twin planet. It is, after the Moon, our closest neighbor, coming within 26 million miles (42 million km) at its nearest. Since Venus orbits closer to the Sun than the Earth by at least this distance, it receives much more heat. This extra heat, combined with the fact that Venus has a deep, mainly carbon dioxide atmosphere which creates a powerful greenhouse effect, makes it too hot for water to be liquid at the surface. All Venus's water is thus in the form of huge clouds cloaking the planet's features. The atmospheric pressure on the surface of Venus is nearly one hundred times what we experience on Earth – it is equivalent to the pressure 3,300 feet (1,000 m) beneath the level of our sea.

*Sulfuric acid clouds on Venus are made by reactions (**right**) in the atmosphere and the planet's crust. There is a wind circulation system driven by the difference in temperature between the poles and the equator, which receives more heat from the Sun.*

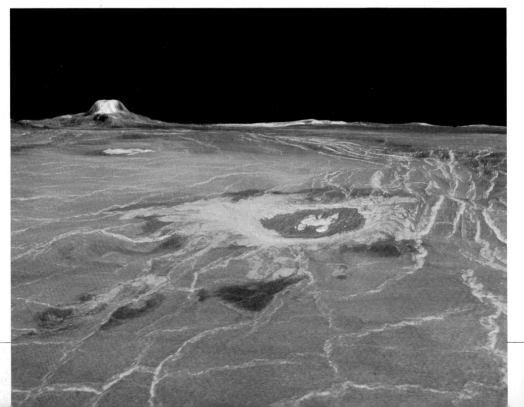

Venus is volcanically active and heat from below the crust vents through volcanoes which belch out vast amounts of sulfur dioxide into the atmosphere, forming deadly clouds of sulfuric acid. The top layers of the clouds whip around the planet at up to 250 mph (400 km/h), so the planet appears to rotate once every four Earth days. Space probes taking radar measurements of the solid surface show that its rotation period is in fact 243 days. Strangely, Venus spins in the opposite direction to most of the other planets. It is thought that this is because the Earth's gravity has locked Venus's rotation to the Earth's orbit, as is indicated by the fact that Venus always shows the same face to Earth when it is closest to us.

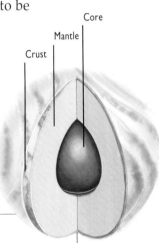

PLANET PROFILE: VENUS

		miles	km
Distance from Sun	max.	67.7 million	109 million
	min.	66.7 million	107.4 million
	ave.	67.2 million	108.2 million
Equatorial diameter		7,520	12,102
Escape velocity		6.437/sec	10.36/sec
Mean orbital velocity		21.77/sec	35.03/sec

Length of year	224.7 Earth days
Rotation period	243.01 Earth days
Surface temperature	869°F 465°C
Atmospheric pressure (at surface)	94 times Earth's
Mass	0.82 of Earth's
Mean density	5.25 g/cm³
Orbital eccentricity	0.007
Angle of equator to orbit	177.3°
Angle of orbit to ecliptic	3.394°
Albedo (reflectivity)	0.64
Gravity at equator	0.88 of Earth's

*The surface of Venus is obscured by clouds (**above**). They reflect sunlight, and astronomers once thought that they would shield the surface below and keep it at a temperature where lush jungle life could exist, as on the Earth's surface 300 million years ago.*

When space probes either landed on or took radar images of the surface, the barren truth was revealed. The Russian Venera *landed in 1970, the American* Pioneer 1 *started a radar survey in 1978, and the* Magellan *orbiter began taking radar pictures of the surface in 1989.*

*The landscape (**left**) is a computer simulation based on radar data sent back to Earth by the* Magellan *orbiter. In the center of the image is the impact crater Maria Cunitz, which is 30 miles (48.5 km) across. On the horizon is the volcano Gula Mons, which is about 2 miles (3 km) high.*

THE ACID CYCLES ON VENUS

A series of complex chemical reactions (**opposite page**) fills the atmosphere of Venus with choking sulfuric acid which falls as rain in the upper atmosphere. There are three principal reactions, or cycles, taking place more or less continuously.

In the fast atmospheric cycle, sulfur dioxide (SO_2) is converted by sunlight into sulfuric acid (H_2SO_4). In the slow atmospheric cycle, hydrogen sulfide (H_2S) and carbon oxysulfide (COS) are converted into H_2SO_4. Both reactions use carbon dioxide and water vapor, present in abundance in the planet's atmosphere. In the lower atmosphere, H_2SO_4 breaks down to form, among other things, sulfur trioxide (SO_3).

In the crustal cycle, iron pyrites (FeS_2) from the planet's surface reacts with water vapor and carbon dioxide to produce H_2S and COS, which go on to react with oxygen to form SO_2. Excess atmospheric SO_2 forms calcium sulfate ($CaSO_4$) which, with iron oxide in the surface and carbon dioxide from the atmosphere, makes FeS_2 once more.

*Like acid rain on Earth, which has eaten away the detail of this carving (**left**) on the exterior of Buckland Abbey in Devon, England, acid rain on Venus interacts with the rocks on the planet. Venus's acid rain, however, is many times more concentrated, though the main component – sulfuric acid formed from sulfur dioxide in the atmosphere – is the same.*

*Radar scans (**below**) reveal that 60 percent of Venus's surface is low-lying plain (colored blue), studded with ancient impact craters. There are two highland regions on the planet (colored green and yellow). One, Terra Ishtar, is the size of Australia. The other, Terra Aphrodite, is the size of northern Africa.*

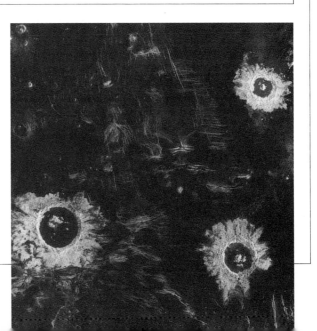

*Three craters – from 22 to 40 miles (35 to 65 km) across – show on this radar image of the low-lying plains of Venus (**right**) taken by the* Magellan *spacecraft in 1990.*

Mars: the red planet

The red glowing light in the sky that is the planet Mars has intrigued humans for thousands of years.

The fourth planet was named after Mars, the Roman god of war, because of its red hue and that color's association with blood. About one-and-a-half times as far away from the Sun as Earth is, Mars receives less than half the heat that Earth does. But it is the only planet to have an atmosphere or day-time temperatures anything like those of the Earth.

The diameter of Mars is just over half that of Earth, but it has only one-tenth of Earth's mass, giving it a surface gravity around a third as strong. It rotates on its axis once every 24.623 hours, giving it a day length about 40 minutes longer than Earth's. But its year is nearly twice as long, since the planet takes 686.98

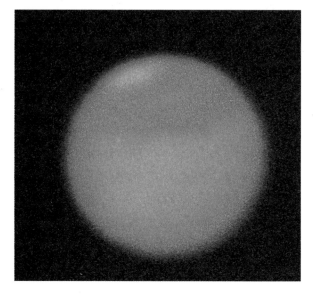

Mantle
Core
Crust

*Mars is properly visible from Earth for only a few months every 26 months, when the planets pass within 35 million and 62 million miles (56 million and 100 million km) of each other. When Mars is close to Earth, it is possible, using a telescope, to see surface features (**above**). Space-probe pictures such as this one (**right**) taken by the* Viking 1 *orbiter on June 18, 1976, provide images far superior to those taken from Earth.*

PLANET PROFILE: MARS

		miles	km
Distance from Sun	max.	154.8 million	249.1 million
	min.	128.4 million	206.7 million
	ave.	141.6 million	227.9 million
Equatorial diameter		4,217	6,786
Escape velocity		3.1/sec	5/sec
Mean orbital velocity		15/sec	24.13/sec
Length of year		686.98 Earth days	
Rotation period		24.62 Earth hours	
Surface temperature	°F	−207 to +72	
	°C	−133 to +22	
Atmospheric pressure	(at surface)	0.007 of Earth's	
Mass		0.11 of Earth's	
Mean density		3.95 g/cm³	
Orbital eccentricity		0.093	
Angle of equator to orbit		25.19°	
Angle of orbit to ecliptic		1.850°	
Albedo (reflectivity)		0.154	
Gravity at equator		0.38 of Earth's	

CANALS: NOW YOU SEE THEM, NOW YOU DON'T

Controversy about the possibility of intelligent life on Mars was sparked in 1877, when Italian astronomer Giovanni Virginio Schiaparelli observed what he thought were straight channels – or, in Italian, *canali*. American astronomer Percival Lowell (1855–1916) (**left**) thought the markings were bands of vegetation bordering canals built by intelligent beings to carry water from the poles. But in the 1950s, large telescopes showed that these canals did not exist. They were made up of smaller, distinct features that Lowell's telescope could not resolve.

*The brain constantly tries to make sense of what the eye perceives. Presented with a row of poorly defined spots, such as separate marks on a planet's surface seen dimly through a telescope, it will interpret a straight line where none exists. Such perceptions led astronomers to construct elaborate maps (**right**) of canals on Mars where there were none.*

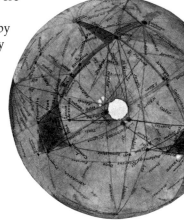

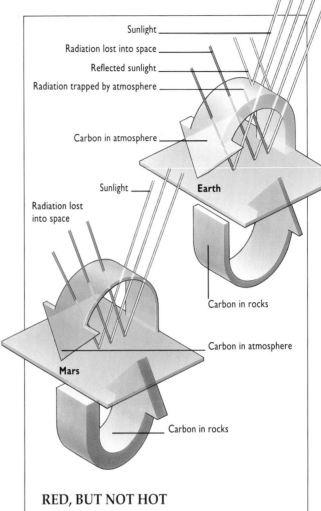

Sunlight

Radiation lost into space

Reflected sunlight

Radiation trapped by atmosphere

Carbon in atmosphere

Earth

Sunlight

Radiation lost into space

Carbon in rocks

Carbon in atmosphere

Mars

Carbon in rocks

Earth days to orbit the Sun. Mars's axis of rotation is tilted at 25.19 degrees to the plane of its orbit, so, like Earth, the planet has seasons. But because Mars's orbit is more eccentric – less circular – than Earth's, the lengths of its seasons are less equal than Earth's. In Mars's northern hemisphere, for example, spring lasts 199 Earth days and is 52 days longer than fall.

The seasons on Mars can be seen in changes to its ice caps, which are visible through a small telescope. Each of Mars's polar ice caps has a central region, made of water ice, which is about 60 miles (100 km) across. This region is surrounded by frozen carbon dioxide – dry ice – which expands and contracts with the seasons. It descends to 60 to 65 degrees latitude in winter, rising above 80 degrees in summer.

RED, BUT NOT HOT

The Earth owes its warmth to the greenhouse effect and to its position closer to the Sun than Mars. Carbon dioxide (CO_2) – among other gases – in its atmosphere traps the energy in sunlight, preventing it from being re-radiated out into space. Volcanic activity releases the CO_2, while biological activity and weathering absorb excess CO_2 to balance the system.

On Mars, atmospheric pressure is less than 1 percent of that on Earth and, despite consisting almost entirely of CO_2, the atmosphere is too thin to retain much of the reduced heat from the Sun. Nor is there any volcanic activity to release CO_2 from the rocks. Temperatures on the equator of Mars thus vary wildly from a warm 72°F (22°C) in the day to an arctic –94°F (–70°C) at night; at the Martian poles, temperatures can drop right down to –207°F (–133°C).

Life on Mars

Astronomers dream there might be life on the red planet – could they be right?

In 1969, the *Mariner 6* and *7* spacecraft photographed the surface of Mars from several hundred miles away. Their pictures showed a barren and pitted surface much like that of the Moon.

Nevertheless, NASA planned to send the Viking probes to Mars and called in British scientist James Lovelock to help design experiments that could determine whether there was life on Mars. Lovelock concluded that it was possible to tell if there was life on a planet simply by analyzing its atmosphere, and Mars's inert, 95 percent carbon dioxide atmosphere showed there was no life.

While it was accepted that large organisms such as animals and plants could not live on Mars, some scientists still believed that microorganisms could exist there. After Earth, Mars is the planet with the most favorable conditions for the emergence of life. The winding valleys seen on the surface are dried-up riverbeds, and the presence

of argon in the atmosphere indicates that it was once much denser, allowing water to remain in its liquid state there.

In 1976, two Viking landers armed with life detectors landed on the surface of Mars. The experiments on board the craft were designed to detect signs of even the most rudimentary microbes. *Viking 2* was shut down in 1979. *Viking 1* continued sending back data once a week until the mid-1990s. No signs of life have yet been found.

*Water must have flowed on Mars at some time. The evidence comes from photos of the Valles Marineris (**above**) taken by the Viking 1 orbiter. The Valles Marineris is a system of canyons and cracks of geological origin in the surface of the planet. Some of them show signs of water erosion, with clearly marked tributaries and winding, dried-up riverbeds. No one is certain when the water was there, but it must have been many millions of years ago.*

*Mars's characteristic color is primarily due to the high content of oxidized (rusted) iron in its surface layers. Like a rusty iron nail (**above right**), Mars is red. The surface is covered with fine dust which is often whipped up into dust storms by winds blowing at up to 200 mph (325 km/h).*

Photographs of the Martian surface sent back by Viking 1, which landed in 1976, show a harsh, rock-strewn terrain under a pink sky. Reddish dust lies between darker bedrock material, and boulders litter the desolate scene.

Should manned missions reach Mars, the first settlers will have to live below ground to protect themselves from cosmic rays. Plants for food could be grown in greenhouses, and water could be melted from the poles.

THE NEW MARTIANS – BRINGING LIFE TO THE RED PLANET

Just because there is no life on Mars now does not mean that there will not be any in the future. There are radical plans to "terraform" Mars, turning the planet into a smaller version of Earth. To raise the temperature, it has been suggested that CFCs and other greenhouse gases – pollutants on Earth – could be released. Detonating nuclear warheads on the surface would also raise the temperature and possibly revive inactive volcanoes, pumping more carbon dioxide into the atmosphere, further adding to the greenhouse effect. Once the frozen water in the polar caps had been melted and a hydrological cycle begun, the planet's surface could be seeded with plants that would gradually alter the balance of the atmosphere, producing oxygen. Slowly, a basic ecosystem could be established.

Olympus Mons (right) is a huge, extinct volcano on Mars. It is an incredible 375 miles (600 km) across. At 15 miles (24 km) high, it dwarfs Earth's largest volcano, Mauna Kea in Hawaii, which measures only 6 miles (9.7 km) from its base on the sea floor to its summit.

The Earth's moon, however, dwarfs Mars's two tiny potato-shaped moons – Phobos and Deimos – which orbit the planet in the sky above this imposing volcano. Phobos is 17 x 14 x 12 miles (27 x 22 x 19 km) and rises and sets twice in a Martian day. It orbits at just 3,730 miles (6,000 km) above the surface. Astronomers predict that it will crash into the planet in less than 100 million years. Deimos is smaller, at 9 x 7½ x 7 miles (15 x 12 x 11 km). It orbits more than 12,500 miles (20,000 km) from Mars.

Jupiter: king planet

By far the largest planet in the solar system, Jupiter is also one of the brightest objects in the sky.

More massive than all the other planets combined, Jupiter, with Mars, is the fourth brightest object in the sky after the Sun, the Moon, and Venus. The large surface area of its disk and its high albedo of 0.42 (the proportion of the light shone on it that it reflects) mean it can reach magnitude –2.8 compared with Venus's brightness of –4.4.

Through even a small telescope, such as a 3-inch (75-mm) refractor, Jupiter is a fascinating sight. It shows as a yellowish disk with darker bands, and its four brightest moons can be seen changing position from night to night as they orbit

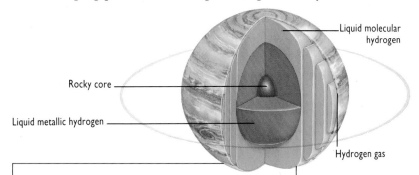

Liquid molecular hydrogen

Rocky core

Liquid metallic hydrogen

Hydrogen gas

PLANET PROFILE: JUPITER

		miles	km
Distance from Sun	max.	506.9 million	815.7 million
	min.	460.4 million	740.9 million
	ave.	483.6 million	778.3 million
Equatorial diameter		88,846	142,984
Polar diameter		83,082	133,708
Escape velocity		37/sec	59.6/sec
Mean orbital velocity		8.11/sec	13.1/sec
Length of year		11.86 Earth years	
Rotation period		9.841 Earth hours	
Surface temperature		–238°F	–150°C
Mass		317.94 times Earth's	
Mean density		1.33 g/cm³	
Orbital eccentricity		0.048	
Angle of equator to orbit		3.1°	
Angle of orbit to ecliptic		1.308°	
Albedo (reflectivity)		0.42	
Gravity at equator		2.34 times Earth's	

around it. Galileo saw this movement of the moons in 1610, and it helped him prove the heliocentric, or Sun-centered, theory of the solar system.

Jupiter is unlike Earth and the other rocky, terrestrial planets – Mercury, Mars, and Venus – which have well-defined, solid surfaces and thin atmospheres above them. On Jupiter and the other gas giants – Saturn, Uranus, and Neptune – the atmosphere makes up virtually the whole bulk of the planet, to the extent that Jupiter's rocky core of about 18,650 miles (30,000 km) diameter is tiny compared with the depth of its atmosphere. Like a star, Jupiter's atmosphere is composed largely of hydrogen, with a lesser amount of helium. There are also traces of methane and ammonia and a tiny amount

The Voyager missions observed Jupiter close up in 1979. They sent back a wealth of images of this majestic planet and its retinue of 16 moons.

Many discoveries were made by the two Voyagers, including the fact that Jupiter has a ring, like Saturn. The ring lies in the planet's equatorial plane at an altitude of about 35,420 miles (57,000 km) above the cloud tops. It is 4,040 miles (6,500 km) wide, less than ½ mile (1 km) thick and probably made of dust and ice.

of water vapor. The pressure in the gas at the surface of the rocky core is 45 million times that on Earth at sea level. This immense pressure has compressed the gas so that it has some of the properties of a solid, and it has also raised the temperature to 30,000K. Heat from the core reaches the surface through convection, and Jupiter radiates 2½ times the amount of energy it receives from the Sun. This energy is given off mostly at infrared wavelengths.

Until the 1940s, it was thought that Jupiter radiated energy like a star. But Jupiter is only about 0.001 times the mass of the Sun. Since the critical mass needed to set off the nuclear reaction that powers stars is about 0.06 times that of the Sun, Jupiter would have to be 60 times bigger to become a star. So the light we see from Jupiter is sunlight reflected by the clouds, not energy from fusion in its interior.

Like a pizza base being spun in the air to make it stretch, Jupiter rotates so fast that it expands at the equator where the planet is more than 5,600 miles (9,000 km) wider than it is from pole to pole.

An image through an amateur telescope (above) shows Jupiter's cloud belts which exist because bands in the planet's gaseous surface have different rotation periods. Those at the equator take about five minutes less to rotate than those nearer the poles. The difference in speed between these belts creates huge eddies that constantly form and re-form. One eddy – the Great Red Spot – was first seen through a telescope in the middle of the 17th century and has been present ever since.

The gas giant

Jovian wonders include a metal-like gas and a storm that could swallow Earth.

When we look at Jupiter, we see its outer atmosphere. This is mainly hydrogen gas with traces of other chemicals hanging in it in layers, like clouds floating in the sky. As the planet rotates, these layers mix up and create banded features which are parallel to the planet's equator. Some 19 separate bands can be observed, and they have been given names according to where they occur on the planet. Winds in these bands range from westerlies blowing at more than 110 mph (180 km/h) to easterlies blowing at up to 270 mph (430 km/h). Where these winds meet, there are violent eddies, and here huge twisted structures evolve.

The North Polar Region shows dark belts and pale areas, while the Northern Temperate Band is fringed with huge red eddies half the size of the Earth. Below that is the bright North Tropical Zone, whose high clouds are made of ammonia crystals. The North Equatorial Belt has a twisted, braided structure caused by complex systems of winds blowing in opposing directions.

Below the equator, the planet's structure becomes more complex. The Great Red Spot (GRS) lies across the dark South Equatorial Belt and the lighter South Tropical Zone. Although the GRS sometimes fades, it never completely disappears and seems to be a permanent feature of the planet's surface – it has been observed for over 300 years. The South Polar Region has enormous storms which appear as huge white ovals of cloud.

The Jovian atmosphere, with all of its visible beltlike features, is about 600 miles (1,000 km) thick. Below this thick atmosphere is an ocean of liquid hydrogen

whose surface is at 2,000K. It does not boil away because the surface pressure is 90,000 atmospheres (90,000 times the average pressure at sea level on Earth).

At the bottom of the ocean, at a depth of approximately 15,500 miles (25,000 km), the temperature is 11,000K, and the pressure is 3 million Earth atmospheres. In these fearsome conditions, hydrogen becomes metallic. Its atoms are pressed so closely together that the electrons are stripped off to run freely through the hydrogen as they do in the crystalline structure of a metal. Huge electric currents running through this metallic shell, 18,650 miles (30,000 km) thick, are responsible for Jupiter's magnetic field, which is 1,000 times larger than Earth's.

Inside the metallic hydrogen is Jupiter's dense, solid core, made of iron and silicates, which is about eight times Earth's volume.

The Great Red Spot (GRS) is a swirling storm hanging in Jupiter's atmosphere lying across two bands in the southern hemisphere. It is about 25,000 miles (40,000 km) across – over three times the diameter of the Earth – and rotates counter clockwise with a period of around six days. The color of the GRS is probably due to phosphine dredged up from lower layers, which is then broken down by sunlight to create red sulfur.

Also suspended in Jupiter's stupendous gas atmosphere are layers of water droplets, ice crystals, and crystals of ammonia and ammonium hydrosulfide (right). These layers have different colors: ammonia crystals are white; ammonium hydrosulfide is reddish; and mixtures of sulfur compounds and water are brown.

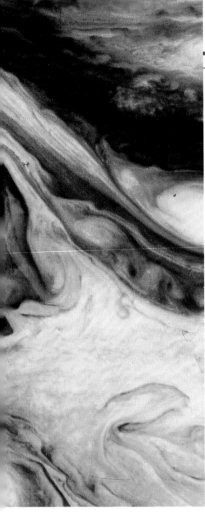

ROBERT HOOKE

British physicist and inventor Robert Hooke (1635–1703), discoverer of the law of elasticity (Hooke's Law), contributed to many areas of science. He made improvements to microscopes and telescopes (**right**) and proposed that Jupiter rotated. In 1664 he discovered the Great Red Spot of Jupiter.

He was the first person to describe the crystalline structure of snowflakes; he discovered plant cells while examining a piece of cork under a microscope; and he was one of the first to propose the theory of evolution after examining microscopic fossils.

Jupiter's Great Red Spot is some 5 miles (8 km) higher than the clouds. At the spot's center, gas spirals up, raising pressure in a dome shape. The spot is 4°F (2°C) cooler than its surroundings.

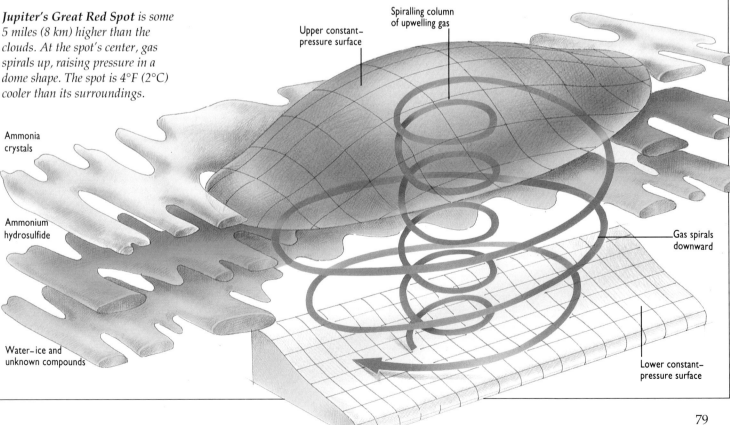

Ammonia crystals

Ammonium hydrosulfide

Water-ice and unknown compounds

Upper constant-pressure surface

Spiralling column of upwelling gas

Gas spirals downward

Lower constant-pressure surface

The king's courtiers

Mighty Jupiter has several large satellites in tow, an extensive magnetic field, and even a ring system.

As befits a mighty planet, Jupiter has at least 16 moons. The four largest – Io, Europa, Ganymede, and Callisto – are called the Galilean moons, as they were discovered by Galileo in 1610. They are a respectable size: both Callisto and Ganymede – the outer Galilean moons – are larger than Mercury.

Within Io's orbit is another small moon, Amalthea. It is a reddish irregular body, 168 by 96 miles (270 by 155 km), with its long axis pointing toward Jupiter. Infrared surveys show it is heated by electric currents running through Jupiter's ionosphere.

There are two other moons even closer to Jupiter. Both are about 25 miles (40 km) in diameter and orbit 36,000 miles (58,000 km) or so above the cloud tops. A third small satellite orbits between Amalthea and Io.

Even closer to the planet are its rings, discovered by *Voyager 1* on March 4, 1979. Until that time, the presence of rings was not certain, but a blurry photo taken by *Voyager* confirmed their existence. The rings are about 35,000 miles (57,000 km) above the tops of Jupiter's clouds and are thought to be formed of dust and ice thrown up by micrometeors striking Jupiter's innermost moons.

Much farther out are the remaining eight known moons. They are probably captured asteroids and form two distinct groups. Four orbit at 7.1 million miles (11.5 million km) and their orbits are all inclined at approximately 28 degrees. The second group is between 12.4 million and 14.9 million miles (20 million and 24 million km) and is inclined at 150 degrees. They range in diameter from 6 to 18½ miles (10 to 30 km).

Ice-covered Callisto (above) is second largest of Jupiter's moons at 3,000 miles (4,825 km) across. It is studded with impact craters.

Europa (right) is also ice-covered, but has fewer craters and at 1,950 miles (3,138 km) across is the smallest Galilean moon. It is believed that water has flowed out from its interior and frozen, resurfacing Europa with fresh ice to make it the smoothest body in the solar system.

Callisto

Europa

Elara
Himalia
Lysithea
Leda

Carme
Pasiphaë
Ananke
Sinope

Amalthea
Adrastea
Metis

Callisto
Ganymede
Europa
Io
Thebe

Magnetopause

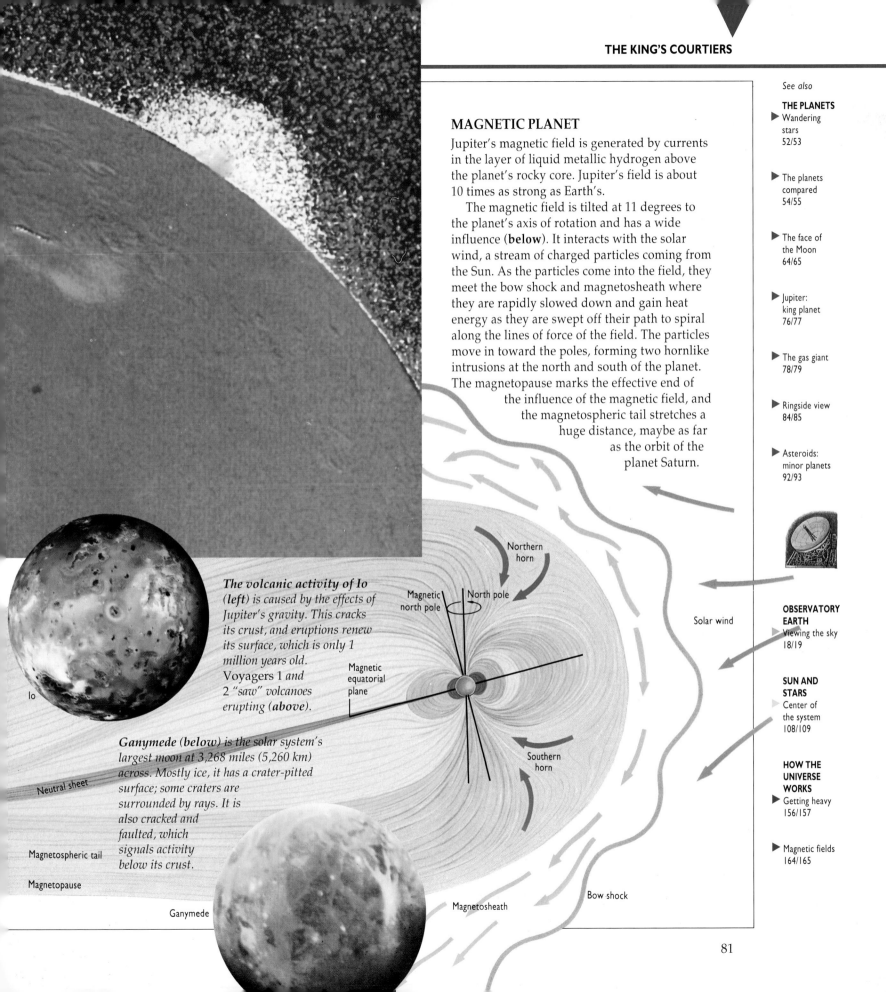

MAGNETIC PLANET

Jupiter's magnetic field is generated by currents in the layer of liquid metallic hydrogen above the planet's rocky core. Jupiter's field is about 10 times as strong as Earth's.

The magnetic field is tilted at 11 degrees to the planet's axis of rotation and has a wide influence (**below**). It interacts with the solar wind, a stream of charged particles coming from the Sun. As the particles come into the field, they meet the bow shock and magnetosheath where they are rapidly slowed down and gain heat energy as they are swept off their path to spiral along the lines of force of the field. The particles move in toward the poles, forming two hornlike intrusions at the north and south of the planet. The magnetopause marks the effective end of the influence of the magnetic field, and the magnetospheric tail stretches a huge distance, maybe as far as the orbit of the planet Saturn.

The volcanic activity of Io (**left**) *is caused by the effects of Jupiter's gravity. This cracks its crust, and eruptions renew its surface, which is only 1 million years old.* Voyagers 1 *and* 2 *"saw" volcanoes erupting* (**above**).

Ganymede (**below**) *is the solar system's largest moon at 3,268 miles (5,260 km) across. Mostly ice, it has a crater-pitted surface; some craters are surrounded by rays. It is also cracked and faulted, which signals activity below its crust.*

Io

Neutral sheet

Magnetospheric tail

Magnetopause

Ganymede

Magnetosheath

Bow shock

Northern horn

Magnetic north pole

North pole

Magnetic equatorial plane

Southern horn

Solar wind

Saturn: lord of the rings

Ringed planet Saturn is a fascinating sight through a small telescope. It also has some unusual properties.

The gas giant Saturn is the second largest planet in the solar system after its big brother Jupiter. It has a little less than a third of the mass of Jupiter and a little more than half its volume. Saturn is the least dense planet – even less dense than water – and, given a bath big enough to put it in, would float.

Saturn spins more slowly than Jupiter, but is slightly more flattened by the rotation because it is less massive. The rocky core is smaller; so is the shell of metallic hydrogen. The planet appears solid, but the disk seen is the tops of the clouds floating in its deep atmosphere – the solid surface of the rocky core is far below the cloud tops.

The clouds have regular banding and a variety of features among them. In the southern hemisphere, there is a red spot 3,725 miles (6,000 km) across, less than a quarter the diameter of Jupiter's Great Red Spot. Saturn also has other permanent, oval features as well as a dark wavy line in the clouds. Its winds are faster than Jupiter's, at up to 1,120 mph (1,800 km/h).

Saturn has a powerful magnetic field, and aurorae have been detected in the polar regions, where charged particles from the Sun are deflected by the field to electrify (ionize) the atmosphere. The planet also generates radio signals, including those characteristic of lightning discharges.

1995

1997

2000

Circulation cells of hydrogen and helium gas

Metallic hydrogen and helium droplets

Rocky core

Rings

Encke's division

Cassini's division

PLANET PROFILE: SATURN

		miles	km
Distance from Sun	max.	936 million	1,507 million
	min.	837 million	1,347 million
	ave.	887 million	1,427 million
Equatorial diameter		74,898	120,536
Polar diameter		67,560	108,728
Escape velocity		20.1/sec	32.3/sec
Mean orbital velocity		6/sec	9.6/sec
Length of year		29.46 Earth years	
Rotation period		10.23 Earth hours	
Surface temperature		–292°F	–180°C
Mass		95.18 times Earth's	
Mean density		0.69 g/cm³	
Orbital eccentricity		0.056	
Angle of equator to orbit		26.7°	
Angle of orbit to ecliptic		2.488°	
Albedo (reflectivity)		0.41	
Gravity at equator		0.93 of Earth's	

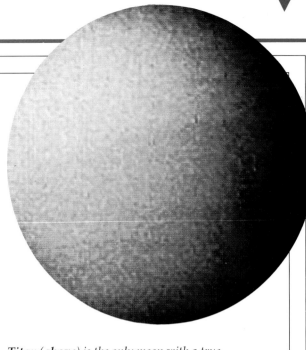

THE MOONS OF SATURN

Saturn has more than 20 satellites. Biggest is Titan (**right**), the solar system's second largest moon, at 3,200 miles (5,150 km) across. Tethys and Mimas have huge impact craters. The Odysseus crater on

The rings of Saturn appear to "open" and "close" over the course of the planet's 24-year orbit. This is because the axis of Saturn (and its rings) is inclined at an angle to the planet's orbit around the ecliptic. The rings are extremely thin, and when seen sideways from Earth they disappear.

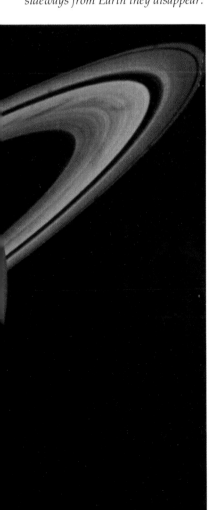

Tethys is 250 miles (400 km) wide – on a moon a mere 652 miles (1,050 km) across. The impact may have shattered Tethys, which re-formed with a giant crack on the side opposite the crater in the form of a huge canyon 3 miles (5 km) deep, 62 miles (100 km) wide, and 1,243 miles (2,000 km) long.

Mimas is 242 miles (390 km) across, yet has an impact crater 84 miles (135 km) wide. It also has crust cracks. Impacts like those on Tethys and Mimas probably threw off debris which added to Saturn's rings.

Several of Saturn's outer moons share an orbit. It is thought that they may once have been one body that broke up. Phoebe, the outermost moon, orbits in the opposite direction to all the rest.

Titan (above) is the only moon with a true atmosphere, composed mainly of nitrogen, with small amounts of methane and cyanide. The conditions on Titan are probably like those on the planets shortly after they formed, and it is thought that chemical reactions there might result in the compounds that are the forerunners of life.

Voyager 1 flew within 4,000 miles (6,500 km) of Titan in 1980. Data sent back shows that the atmospheric pressure on the surface is about twice that at sea level on Earth. The surface of Titan is probably covered by huge oceans of liquid methane. These are obscured from direct view, however, by an orange haze of what on Earth would be thought of as photochemical smog.

2003

2005

2007

2009

2011

2013

2016

2018

2021

Ringside view

The rings of Saturn are one of the wonders of the solar system. And space probes have revealed mysterious happenings within them.

Like so many phenomena in the solar system, the rings of Saturn were first seen by Galileo. In 1610, he perceived two bright spots on either side of the planet. Although his telescope was powerful enough to see the disk of Saturn, it was not good enough to resolve the rings, and he interpreted what he saw as a triple planet.

Nearly half a century later, Christiaan Huygens noted that Saturn was "surrounded by a thin ring not adhering to the planet at any point and inclined to the ecliptic." Huygens thought the ring was solid.

In 1675, Italian astronomer Jean Dominique Cassini, working in Paris, discovered that what had seemed to be a single ring was actually split in two – now called the A and B rings. The gap between these rings is named

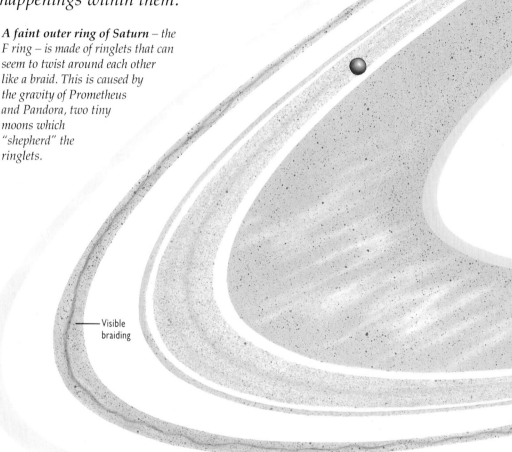

A faint outer ring of Saturn – the F ring – is made of ringlets that can seem to twist around each other like a braid. This is caused by the gravity of Prometheus and Pandora, two tiny moons which "shepherd" the ringlets.

Visible braiding

The broadest, brightest ring, the B ring, has dark spokes that run across it like the spokes in an old-time wagon wheel (left). The spokes, which come and go, seem to be formed by millions of minute, electrically charged particles, each a thousandth of a millimeter across, suspended above the plane of the ring. Saturn's magnetic field makes the particles align into spokes, revolve with the planet for a revolution or two, then vanish.

the Cassini division. A third, darker ring – the C, or *crêpe*, ring – was detected closer to the planet in 1850.

James Clark Maxwell showed, in 1856, that a solid ring would be torn apart by Saturn's gravity field. He was right. Saturn's rings are, in fact, narrow bands of fine debris.

Probes sent to the planet recently have discovered more rings and revealed much detail. Moving out from the planet, first comes the faint D ring, then the less faint C ring, then the bright B ring. Next is the Cassini

Ice and dust particles

Each of Saturn's rings is no more than a few tens of yards thick and is made up of countless millions of blocks of water ice, some dust, and possibly some metallic material. The size of the particles ranges from that of ice cubes to chunks as large as a refrigerator or slightly bigger. Each particle is, in effect, a tiny satellite of the planet and each has its own orbit albeit confined within its ring or ringlet.

The rings start 4,350 miles (7,000 km) above Saturn's clouds and stretch out more than 46,000 miles (74,000 km) into space. The rings are so thin in relation to their width that they are like a huge sheet of tissue paper spread over the area of a football field.

Each ring is made up of many ringlets, and it is thought that the gravitational effect of small moons orbiting within the rings shepherds particles into these rings-within-rings. As a moonlet, or shepherd moon, moves through a ringlet (**below**), an inner particle within a ring is decelerated and drops to an orbit nearer Saturn. A particle farther out is accelerated and boosted into a higher orbit. Thus particles are swept out of certain orbits and concentrated in bands.

Inner particle

Outer particle

Shepherd moon

division – now known to be filled with faint ringlets. The less bright A ring with its own gap, the Encke division, follows. Beyond the A ring come three faint rings: the narrow, often braided F ring; the tenuous G ring; and last, the E ring. Most rings are divided into ringlets, some only tens of yards wide – there are probably 10,000 or more of these.

Astronomers have long puzzled over the origin of the rings. One widely accepted theory says that the rings may be fragments of a moon that could never form because its constituents orbited so close to the planet that gravitational forces tore up any large clumps of matter.

CHRISTIAAN HUYGENS

Dutch astronomer, physicist, and mathematician Christiaan Huygens (1629–95) not only discovered the rings of Saturn, but also propounded a wave theory of light and made significant contributions to the science of mechanics.

He developed a way of grinding and polishing glass that improved telescope lenses. The telescope he constructed with the new lenses gave much better views and it enabled him to discover a satellite of Saturn in 1655. The next year, he saw the stellar components of the nebula in Orion and in 1659 he saw Saturn's rings.

His interest in astronomy and his consequent need to have accurate timekeeping devices led him to invent the clock pendulum. Huygens's interests were extremely wide-ranging and he visited London in 1689 to lecture on his theory of gravitation. He met and corresponded with Isaac Newton. Newton's theory of gravitation, however, trounced that of Huygens, but wave theories of light routed Newton's corpuscular theory.

Uranus

A hazy blue, distant world shining dimly far out from the Sun, Uranus is an enthralling planet with a weird axial tilt.

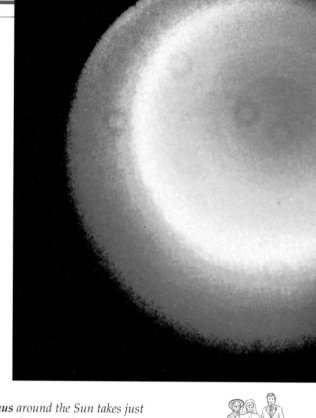

Like Saturn and Jupiter, Uranus is a gas giant. Its diameter is about four times that of Earth, and its central rocky core is about the same size as Earth.

The extraordinary thing about Uranus is its axis, which is tilted so far over that its north pole is 8 degrees below the plane of its orbit. This means that the planet rotates on its axis in a retrograde, or backward, direction compared with most of the other planets. The Uranian seasons are exaggerated by this tilt. Each pole receives 42 years of continuous sunlight, followed by 42 years of darkness.

How the tilt came about is not known, but it has been calculated that it would have taken a collision with a

Uranus (right) is shrouded in a haze of methane gas present in the planet's thick atmosphere. This haze obscures the layers below, and even images from Voyager 2 in visible light show little besides the planet's blue tinge.

Infrared and radar surveys, however, show that the upper cloud layer of Uranus is banded with weather systems like those seen on Jupiter and Saturn. These systems rotate more slowly than the solid core of the planet.

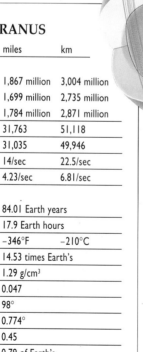

Atmosphere of hydrogen, helium, methane

Rocky core

Mantle of water, methane, ammonia ices

PLANET PROFILE: URANUS

		miles	km
Distance from Sun			
	max.	1,867 million	3,004 million
	min.	1,699 million	2,735 million
	ave.	1,784 million	2,871 million
Equatorial diameter		31,763	51,118
Polar diameter		31,035	49,946
Escape velocity		14/sec	22.5/sec
Mean orbital velocity		4.23/sec	6.81/sec

Length of year	84.01 Earth years
Rotation period	17.9 Earth hours
Surface temperature	−346°F −210°C
Mass	14.53 times Earth's
Mean density	1.29 g/cm³
Orbital eccentricity	0.047
Angle of equator to orbit	98°
Angle of orbit to ecliptic	0.774°
Albedo (reflectivity)	0.45
Gravity at equator	0.79 of Earth's

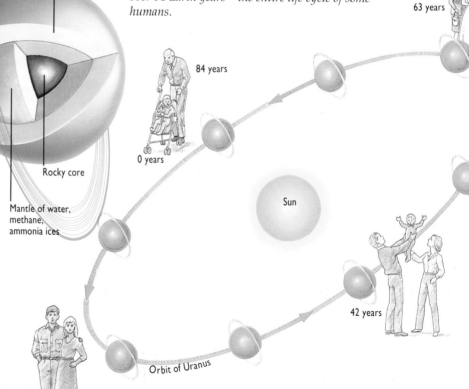

The orbit of Uranus around the Sun takes just over 84 Earth years – the entire life cycle of some humans.

63 years

84 years

0 years

Sun

42 years

Orbit of Uranus

21 years

body about half the size of the Earth to knock the axis into its current position.

Although Uranus has a deep atmosphere, which creates a pressure at the center of its rocky core 20 million times that of Earth's atmosphere at sea level, the planet is too small to have metallic hydrogen around its core like Jupiter and Saturn. But it does have a magnetic field, perhaps generated by currents in its core. Its magnetic field is wildly tilted, too, with the magnetic poles 60 degrees from the geographic poles. This arrangement acts as a magnetic rotor, turning the tail of the magnetosphere into a huge corkscrew.

Uranus has a system of nine rings, which lie between about 10,000 and 16,000 miles (16,000 and 26,000 km) above the tops of the planet's clouds. Although they are not as spectacular as Saturn's, Uranus's rings are unusual because they contain the darkest material yet discovered in the solar system. Eight of the rings are less than 6 miles (10 km) wide, although this does vary. Epsilon, the outermost ring, is between 19 and 62 miles (30 and 100 km) across. Some of the rings are slightly off the equatorial plane, which indicates that they may have been formed comparatively recently.

WILLIAM HERSCHEL

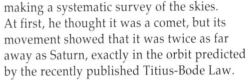

Uranus was seen on several occasions, but was not recognized as a planet until 1781, when British amateur astronomer William Herschel (1738–1822) noticed its movement while making a systematic survey of the skies. At first, he thought it was a comet, but its movement showed that it was twice as far away as Saturn, exactly in the orbit predicted by the recently published Titius-Bode Law.

He named it after George III; the French named it Herschel after the man himself, but German astronomer Bode suggested naming it Uranus – Saturn's mythological father. The discovery made Herschel famous, and he was able to give up his job as a musician and devote himself to astronomy – many consider him the founder of modern stellar astronomy.

THE FRACTURED MOON

Miranda (**right**), the smallest and closest of Uranus's five major satellites, seems to have been blasted apart by a huge impact and then reassembled. Its surface is a jumble of cliffs, craters, and canyons – some ten times deeper than Earth's Grand Canyon, though Miranda is only 310 miles (500 km) across.

Oberon, the outermost moon, has many craters, some of which show signs of volcanic activity. The largest moon is Titania at 1,000 miles (1,600 km) across. It is also heavily cratered and has a valley 930 miles (1,500 km) long and 47 miles (75 km) wide. Ariel's surface is relatively crater free, but it is crisscrossed by grooves up to 19 miles (30 km) deep. It seems that it has been resurfaced by volcanic activity. Although it has a crater 68 miles (110 km) across, Umbriel shows no sign of volcanism. Uranus has at least 10 other moons, and they, like the five above, are made of rock and a mix of ice and methane.

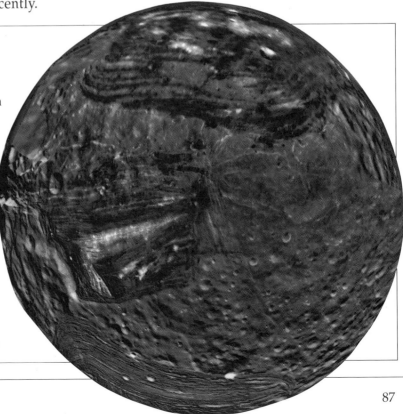

Neptune

Named after the Roman sea god, Neptune is the eighth of the solar system's planets and another gas giant.

In 1820, French astronomer Alexis Bouvard began making tables predicting the movements of Jupiter, Saturn, and Uranus. While the tables worked for Jupiter and Saturn, he found it impossible to predict where Uranus should be. This encouraged the idea that there might be a planet beyond Uranus disturbing the orbit of Uranus with its gravity.

Two mathematicians, John Couch Adams in England and Urbain Le Verrier in France, set about calculating where the missing planet should be. Le Verrier's results were published in France on August 31, 1846. And on September 23, 1846, Johann Galle in Berlin confirmed the existence of a new planet, less than one degree from the predicted position.

Neptune is another gas giant, similar in size and structure to Uranus. Around the core is an ocean of water, methane, and ammonia, thousands of miles in

depth. The deep atmosphere above the ocean is mainly hydrogen and helium, with a little methane.

There are dark spots in the planet's cloud surface. One of these, the Great Dark Spot, is a huge storm system the size of the Earth. The solar system's fastest winds, reaching speeds of 1,250 mph (2,000 km/h), blow around it.

Like other gas giants, Neptune has rings. There are four dim rings: two broad and two narrow. The narrow rings are

PLANET PROFILE: NEPTUNE

		miles	km
Distance from Sun	max.	2,819 million	4,537 million
	min.	2,769 million	4,456 million
	ave.	2,794 million	4,497 million
Equatorial diameter		30,775	49,528
Polar diameter		30,199	48,600
Escape velocity		14.5/sec	23.3/sec
Mean orbital velocity		3.37/sec	5.43/sec
Length of year		164.79 Earth years	
Rotation period		19.2 Earth hours	
Surface temperature		−346°F	−210°C
Mass		17.14 times Earth's	
Mean density		1.64 g/cm³	
Orbital eccentricity		0.009	
Angle of equator to orbit		29.6°	
Angle of orbit to ecliptic		1.774°	
Albedo (reflectivity)		0.50	
Gravity at equator		1.12 times Earth's	

Rings

Atmosphere of hydrogen, helium, methane

Mantle of water, methane, ammonia ices

Rocky core

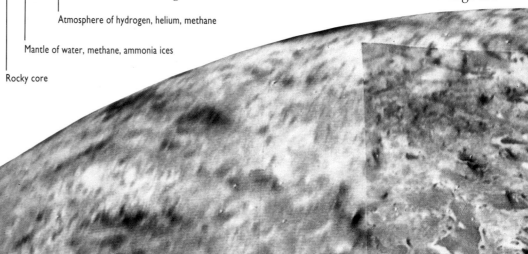

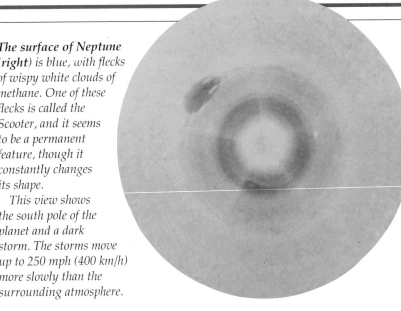

shepherded by two small moons – Galatea and Desponia. Two other small moons – Naiad and Thalassa – orbit within the ring system, but do not seem to act as shepherd moons. They are less than 125 miles (200 km) across and may be fragments of a larger moon that has broken up. Outside Neptune's rings are the small moon Larissa, and beyond it Proteus, which has a crater covering almost half its surface. The outermost moon, Nereid, orbits at 3.42 million miles (5.51 million km) from the planet's surface and is probably a captured asteroid.

Neptune (left) was photographed by the space probe Voyager 2 *in 1989. Among other things, the mission discovered six of the planet's eight known moons. Neptune was the probe's last stop on its trip through the solar system.*

The surface of Neptune (right) is blue, with flecks of wispy white clouds of methane. One of these flecks is called the Scooter, and it seems to be a permanent feature, though it constantly changes its shape.

This view shows the south pole of the planet and a dark storm. The storms move up to 250 mph (400 km/h) more slowly than the surrounding atmosphere.

Triton (below) is the solar system's coldest body. Despite a chilling temperature of −391°F (−235°C), it is volcanically active, and its surface ice is cracked where it has melted and refrozen.

GEYSERS ON TRITON, COFFEE ON EARTH

At the south pole of Triton, Neptune's largest moon, are nitrogen geysers that blast out black dust plumes 5 miles (8 km) high. Winds in the thin nitrogen and methane air carry the dust 95 miles (150 km), streaking Triton's surface with soot.

A geyser works like a coffee percolator (**left**). Pressure builds up in a chamber at the base of a tube filled with liquid: liquid nitrogen on Triton and water in a percolator. When the pressure builds up enough, liquid in the column above the chamber is forced up – to make a geyser plume or to let water trickle through coffee.

- Hot water forced out of tube
- Hot water percolates through coffee
- Bubble of steam forced up tube
- Boiling water
- Bubble of steam builds up in cone

Pluto and Charon

Remote, icy Pluto, locked in an intimate orbit with its moon Charon, was only discovered in 1930 after painstaking detective work.

At the beginning of the 20th century, American astronomer Percival Lowell, among others, made calculations of where a ninth planet might be. The search for this planet was inspired by the fact that the discovery of the eighth planet, Neptune, did not completely account for the irregularities in the orbit of the seventh planet, Uranus.

And observations of Neptune itself showed it did not orbit as expected. Between 1905 and his death in 1916, Lowell and his colleagues made telescopic sweeps of the area where the planet was thought to be, but they were unable to find it.

Over a decade later, Lowell's successors at the Lowell Observatory in Flagstaff, Arizona, installed a new telescope and appointed young astronomer Clyde W. Tombaugh to continue the search. He painstakingly photographed the sky and compared pictures taken several days apart, looking for moving objects. On February 18, 1930, he found something that had moved against the background stars, compared with the plates he had taken on January 21, 23, and 29. It was 5 degrees from where Lowell had predicted. The announcement of the discovery was made on March 13, 1930. The planet was named Pluto, the Roman god of the lower world, whose name happens to begin with the initials of Percival Lowell.

There may be a planet beyond Pluto, since its gravitational pull is not large enough to

PLANET PROFILE: PLUTO

		miles	km
Distance from Sun	max.	4,583 million	7,375 million
	min.	2,939 million	4,730 million
	ave.	3,675 million	5,914 million
Diameter		1,419	2,284
Mean orbital velocity		2.9/sec	4.7/sec
Length of year		248.54 Earth years	
Rotation period		6.39 Earth days	
Surface temperature		−364°F	−220°C
Mass		0.0022 of Earth's	
Mean density		2.03 g/cm³	
Angle of orbit to ecliptic		17.148°	
Orbital eccentricity		0.246	
Gravity at equator		0.04 of Earth's	

PLANET PROFILE: CHARON

		miles	km
Distance from Pluto	ave.	12,204	19,640
Diameter		741	1,192
Mean orbital velocity		2.9/sec	4.7/sec
Time to orbit Pluto		6.39 Earth days	
Rotation period		6.39 Earth days	
Surface temperature		−364°F / −220°C	
Mass		0.0003 of Earth's	
Mean density		2 g/cm³	
Orbital inclination		98.8°	
Orbital eccentricity		0.246	
Gravity at equator		?0.02 of Earth's	

Clearly showing Pluto and Charon, this Hubble Space Telescope photograph (**above**), taken in 1990, reveals the closeness of the orbits and the similarity in brightness of the two bodies. Because of their similar size and mass, they are often considered to be a double planet, rather than a single planet with a satellite.

Orbit of Neptune

Orbit of Pluto and Charon

Diameter of Earth 7,926 miles (12,756 km)

Distance from Pluto to Charon 12,204 miles (19,640 km)

Rocky core

Pluto's orbit is the most oval of those of all the planets – it spends 20 years of its 248-year orbit closer to the Sun than Neptune. The most recent 20-year period began in 1979. Its orbit is tilted at an angle of 17° to the ecliptic – the plane in which the planets orbit the Sun.

Methane frost

Mantle of ices

account for all the perturbations of Neptune. The hunt for a tenth planet is now on.

Pluto is tiny, even compared with Earth, which is five times larger and 500 times as massive. Pluto's rocky core is covered with a thick layer of water ice, overlaid with a layer of ice mixed with frozen methane. When closest to the Sun, a thin atmosphere of methane and, probably, nitrogen boils off the surface. In sunlight, its surface appears pinkish, indicating the presence of carbon, and there are white polar caps, probably made of frozen methane.

In 1978 American astronomer Jim Christy found that Pluto had a satellite. The satellite, Charon, is over half Pluto's diameter and, at 12,204 miles (19,640 km) from it, 20 times closer than our moon is to us. Charon is much like Pluto, but too small to retain methane, so its surface is pure ice. Pluto and Charon are pitted with craters. Both bodies are so small that it is thought they may be asteroids.

SHARED ORBIT

All multibody systems orbit around their center of gravity. Pluto and Charon are so close in size that their center of gravity lies between them outside Pluto.

The center of gravity follows the orbital path, so moon and planet loop along as they orbit each other like an unbalanced baton being twirled by a member of a marching band. Lighter Charon loops more than Pluto.

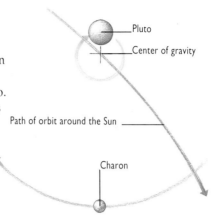

Pluto
Center of gravity
Path of orbit around the Sun
Charon

Sun

Diameter of Moon
2,160 miles (3,476 km)

Charon and Pluto are so close – just the width of Earth plus two Moons apart – that tidal force has locked their spins and orbits. They spin in opposite directions and their same sides always face each other. Seen from Pluto's surface, Charon is four times the diameter of our Moon and hangs motionless in the sky, never rising or setting.

Asteroids: minor planets

*Countless rocky fragments orbit the Sun, mostly in a loose group
between Mars and Jupiter. But some might spell disaster for Earth.*

Between Mars and Jupiter, where it looks
as if there should be a planet, orbit
thousands of rocky bodies – the asteroids.
Early ideas about their origin held that they
were remnants of a planet that had blown up
for some reason. But astronomers now have a
less dramatic theory for their existence.

When the major planets were accumulating
at the birth of the solar system, there was
simply not enough material available
between what became Mars and Jupiter to
make one big planet. The gravity of the
material of what was becoming Jupiter
prevented a single planet from forming.

Today, the largest chunk, known as Ceres,
is a mere 600 miles (1,000 km) across.
The rest are of decreasing size, the
smallest known being only a few tens
of yards across. Astronomers detect
new ones all the time, mostly as
they trail across photographs
taken of the sky for other
reasons. Each asteroid is given a
catalog number, and some are
named for astronomers by the
International Astronomical
Union. For example, asteroid
4024 bears the name Ronan.

Planet	Saturn		Jupiter	Asteroid belt	Mars	Venus	Mercury	Earth
Predicted distance	100		52	28	16	7	4	10
Actual distance		95.4	52	27.7	15.2	7.2	3.9	10

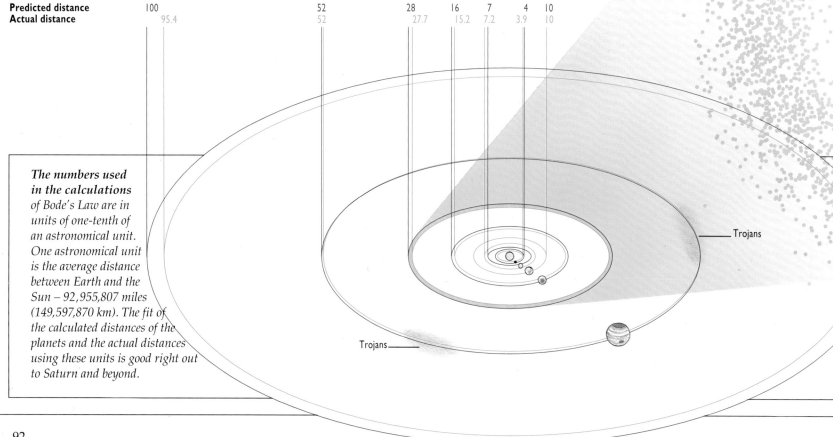

*The numbers used
in the calculations
of Bode's Law are in
units of one-tenth of
an astronomical unit.
One astronomical unit
is the average distance
between Earth and the
Sun – 92,955,807 miles
(149,597,870 km). The fit of
the calculated distances of the
planets and the actual distances
using these units is good right out
to Saturn and beyond.*

Trojans

Trojans

Trojans

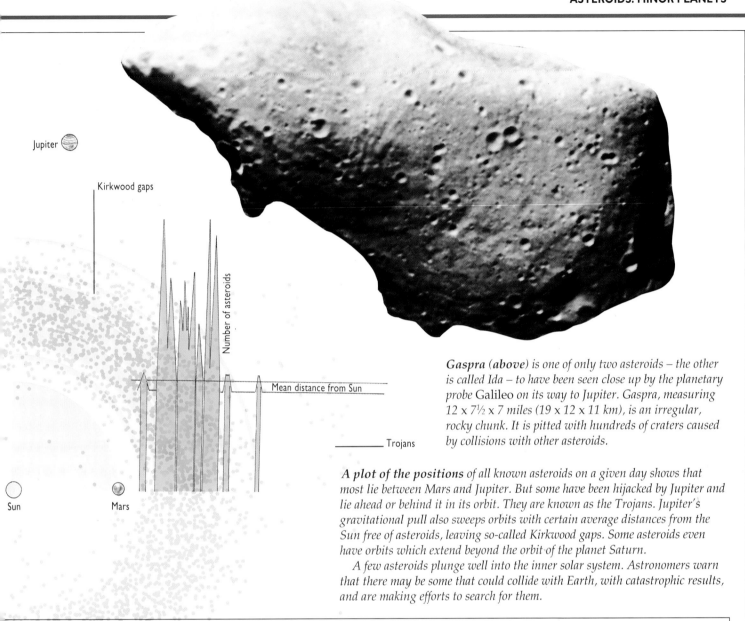

Jupiter

Kirkwood gaps

Number of asteroids

Mean distance from Sun

Sun

Mars

Trojans

Gaspra (above) is one of only two asteroids – the other is called Ida – to have been seen close up by the planetary probe Galileo *on its way to Jupiter. Gaspra, measuring 12 x 7½ x 7 miles (19 x 12 x 11 km), is an irregular, rocky chunk. It is pitted with hundreds of craters caused by collisions with other asteroids.*

A plot of the positions *of all known asteroids on a given day shows that most lie between Mars and Jupiter. But some have been hijacked by Jupiter and lie ahead or behind it in its orbit. They are known as the Trojans. Jupiter's gravitational pull also sweeps orbits with certain average distances from the Sun free of asteroids, leaving so-called Kirkwood gaps. Some asteroids even have orbits which extend beyond the orbit of the planet Saturn.*

A few asteroids plunge well into the inner solar system. Astronomers warn that there may be some that could collide with Earth, with catastrophic results, and are making efforts to search for them.

BODE'S LAW AND THE CELESTIAL POLICE

In 1772 German mathematician Johann Daniel Titius devised a simple number sequence that matched the planets' distances from the Sun. He took the series 0, 3, 6, 12, 24, 48, and 96 in which, except for the first step, each number is double the one before. He then added 4 to each to make 4, 7, 10, 16, 28, 52, and 100. Putting Earth at distance 10, the planets then known fitted the sequence well, and the discovery of Uranus in 1781 – predicted distance 196, actual distance 192 – seemed to confirm its validity.

Because of the sequence's accuracy, Berlin astronomer Johann Bode promoted the idea that

there must be a planet in the gap between Mars and Jupiter at distance 28. Accordingly, a team of planet-hunters, known jocularly as the "celestial police," was assembled to track it down.

But by chance, Italian astronomer Giuseppe Piazzi beat them to it. A "star" that he saw on January 1, 1801, had moved by the next night. This was, in fact, the asteroid Ceres, real distance 27.7. In time, other tiny planets were found, and it was realized that there is no single planet that fills the gap in the sequence now known as Bode's Law or, more fairly, the Titius-Bode Law.

Comets: dirty snowballs

A comet can create a grand spectacle in the heavens, yet the basis of the display is extraordinarily insubstantial.

A prominent comet can look like a dagger, poised among the stars, with a bright head and a long, curved tail. It moves only slowly through the sky from night to night, until it either fades away over several weeks or is lost in the Sun's glare. But despite these occasional fabulous displays, most of the 20 or so comets that appear each year are far too faint to be seen without a telescope, and most do not develop a tail.

Every comet has a nucleus, which is the source of all its activity.

The nucleus is tiny, only a few miles across, and made largely of water ice. But if a comet comes into the inner solar system, within about the orbit of Jupiter, the ice begins to vaporize in the heat of the Sun, giving off gas and dust.

The cloud of gas and dust spreads out from the nucleus and reflects the sunlight so that what was previously a tiny, unnoticed speck suddenly becomes visible. As the comet nears the Sun, the stream of particles and radiation from the Sun sweep the gas and dust away from the nucleus, forming a hazy head, or coma, and sometimes a tail as well. Some of the biggest comets have had tails 100 million miles (160 million km) long – more than the distance from the Earth to the Sun.

Halley's comet (right) returned in 1986, though it was hard to see without optical equipment. During its approach, the spacecraft Giotto went within a few hundred miles of its nucleus and took close-up photos before its camera was wrecked by a collision with cometary material. The pictures showed a chunk of material 10 x 5 miles (16 x 8 km), gushing jets of gas and dust from its sunward side.

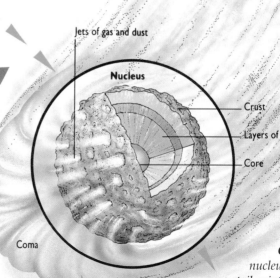

Dust tail

Orbit of comet

A comet's nucleus is covered by a blackish, sootlike layer of dust. The black surface readily absorbs the Sun's heat, which vaporizes the frozen matter beneath, giving off gas and dust (right) to make the glowing coma, or head, which we see.

Jets of gas and dust

Nucleus

Crust

Layers of ice

Core

Coma

In the vacuum of space, ice in a comet's head turns to gas without going through a liquid stage. The same thing happens to dry ice, solid carbon dioxide (**above**).

Solar wind

Gas given off by a comet's nucleus shows as a straight bluish tail pointing directly away from the Sun. Dust forms a curved yellowish tail behind the comet's head.

Some comets are captured by the gravity of major planets, notably Jupiter, and their orbits become less elliptical. These become periodic comets, like Halley's comet which orbits the Sun every 76 years. The periodic comets Tempel-Tuttle and Neujmin 3 also have short, oval orbits.

Gas tail

Neujmin 3
Tempel–Tuttle
Kopff
1910/1986 1985 1984 1983 Halley 1948

Neptune   Uranus   Saturn   Sun

Comets are part of the solar system, but they originate at its very edge and are probably simply icy material mixed with dust – literally "dirty snowballs" – that failed to come together to form true planets. From time to time, an individual comet is diverted in toward the Sun, possibly as a result of the minute gravity effects of other stars. It moves in a long elliptical orbit, taking maybe 100,000 years to reach the inner solar system. When it arrives, its gas and dust are liberated, and a brilliant new comet appears for a few weeks before its orbit takes it away again.

Sometimes a new comet has its orbit changed when it comes close enough to one of the planets to be influenced by its gravity. This can make the orbit smaller, so the comet comes back every few years as a so-called periodic comet. New comets coming close to the Sun have large amounts of dust and can have spectacular tails, while old, periodic comets have lost much of their material after repeated heating by the Sun.

The most famous comet of all is a periodic one called Halley's comet, which spends most of its time quiescent, moving slowly around the far end of its orbit beyond the path of Neptune. Only when it comes into the inner solar system does it put on a show.

Comets are thought to originate in a region far from the Sun called the Oort cloud. The cloud is approximately spherical, and the cold comet nuclei orbit the Sun at all angles to the ecliptic (the plane of the orbits of the planets). So when a comet is directed into the inner solar system, it can approach at any angle. Comet Kopff, for instance, has an elongated elliptical orbit almost at a right angle to the ecliptic.

It is also thought that some short-period comets might come from the Kuiper belt, a region in the plane of the ecliptic extending beyond the orbit of the planet Neptune.

95

Meteors: fires in the sky

A shooting star marks the fiery end of a piece of space debris that has been orbiting the Sun for millions of years.

If you are looking at the night sky, you may occasionally see a brief fiery trail that looks as if a star has suddenly fallen from the sky. The trail is a shooting star, or meteor, and occurs when a particle of matter from space burns up as it hits the Earth's atmosphere. But meteors make up only a tiny fraction of the amount of matter that comes into contact with Earth.

Each day, in fact, something like 1,000 tons of material falls onto the Earth from space. Most of this matter is in the form of minute specks of dust. Although they hurtle into Earth's atmosphere at speeds of up to 212,500 mph (342,000 km/h), most of these tiny particles are so small that they are gently slowed down by friction when they bang into the gas molecules in the upper atmosphere and so leave no fiery trail behind them. When these tiny particles, called micrometeorites, have been slowed by the upper atmosphere, they float down to the Earth's surface. It is only larger particles, bigger than a grain of sand, that burn up with a flash of light.

Whatever their size, most of these particles are the dust from the tails of comets. As a comet moves around the Sun, the debris it leaves behind spreads into a broad band, far wider than its original track. If the Earth passes through one of these clouds of particles, many meteors an hour may be seen

A meteorite (right) shows signs of the intense heat generated by friction with the air as it sped through the atmosphere on its way to the surface.

Some meteorites are small pieces of asteroid whose orbits take them across the path of the Earth around the Sun. But this chunk is thought to have come from Mars after being blasted off the surface when the planet was hit by a large meteorite.

A typical meteor track looks like a streak in the sky (above) and fades almost as soon as it is made. The sort of particle that produces an average meteor may be about the size of a pea, but weighs only a quarter as much – about 0.004 ounces (0.1 g). This suggests that it is not a solid rocky chunk, but is instead a collection of smaller grains, usually dust from a comet. When a pea-sized particle with this structure dashes through the atmosphere, it quickly crumbles and heats up, causing the flare of light.

Before a particle hits the Earth's atmosphere, it is known as a meteoroid, irrespective of its size. If it is a tiny dust speck that hits the atmosphere without leaving a fiery trail, it is called a micrometeorite. If it is large enough to leave a trail it is called a meteor, and if it is larger still and makes it to the Earth's surface relatively intact, it is called a meteorite.

Meteorites also leave trails, and some of them can have spectacular, long-lasting glowing tracks, or even resemble balls of fire.

When a large piece of interplanetary debris hits the Earth, it can make a big crater. Between 30,000 and 50,000 years ago, a body about 100 feet (30 m) across made mostly of iron slammed into the desert of Arizona. The meteorite weighed about 70,000 tons and was traveling at 43,000 mph (69,000 km/h). The force of impact blasted a crater 4,100 feet (1,250 m) wide by 571 feet (174 m) deep, leaving behind Meteor (or Barringer) Crater. If the same thing happened today in a built-up area, it would be as devastating as a nuclear bomb.

in what is called a meteor shower. The meteors all seem to come from the same direction and appear to radiate away from a particular part of the sky. This is just a result of perspective – you see the same effect when you drive into a snowstorm and the snowflakes appear to be radiating out in straight lines from ahead.

Although most meteors are caused by cometary debris, some involve larger pieces of material, most often small fragments of an asteroid. Sometimes the debris does not burn up completely as it passes through the atmosphere and may land on the Earth's surface, where it can be recovered. Then it is known as a meteorite. It is thought that many meteorites hit the surface each year, though it is rare to see this happen. Most that hit are small – up to basketball-sized – but at irregular intervals, much larger pieces hit Earth.

SCATTERED SHOWERS FORECAST

Meteor showers occur when the Earth's orbit around the Sun takes it through a dispersed trail of debris left behind by a comet. A shower is named after the constellation out of which the meteors appear to radiate. Thus to see the Orionids, look at the night sky on the dates shown. Unusually, the Geminids are caused by debris from an asteroid (named Phaethon), not a comet.

Shower name	Duration dates	Date of max. activity	Meteors per hour (max.)	Associated comet
Quadrantids	Jan. 1–6	Jan. 3–4	100	Not known
Lyrids	Apr. 19–24	Apr. 22	10	Thatcher
Eta (η) Aquarids	May 24–Apr. 20	May 4–5	35	Halley
Delta (δ) Aquarids	Aug. 15–July 15	July 28–31	30	Not known
Perseids	Aug. 23–July 30	Aug. 12	80	Swift-Tuttle
Orionids	Oct. 16–27	Oct. 21	25	Halley
Taurids	Nov. 26–Oct. 25	Nov. 3	10	Encke
Leonids	Nov. 15–19	Nov. 17	variable	Tempel-Tuttle
Geminids	Dec. 7–15	Dec. 13	90	(Phaethon)
Ursids	Dec. 17–24	Dec. 22	5	Tuttle

Far out

The solar system stretches out for a long way beyond the orbit of Pluto.

Astronomers have speculated for years that there might be another planet beyond tiny misfit ninth planet Pluto. Measurements of the orbits of Pluto's gas giant neighbors, Uranus and Neptune, suggested that the gravity of another body could be tugging them slightly out of position. However, searches for the tenth planet, known as Planet X, drew a blank until 1992, when an object was found orbiting at about the same distance as Pluto. It was given the designation 1992 QB1, and observations showed that it is a cometlike body 125 miles (200 km) across. In 1993 another object almost as distant was spotted, and several more have since been seen.

But these newly discovered objects – now thought to be part of a band of bodies called the Kuiper belt – are all too small to affect the orbits of the giant planets. Recent studies of the motions of *Pioneer 10*, a space probe that has now left the solar system, and of Halley's comet suggest that there cannot be enough material beyond Pluto to form a planet. It is now thought that errors in observations could account for the apparent changes in the orbits of Uranus and Neptune.

A study of orbits, but this time those of comets, gave astronomers reason to suspect that there might be yet another belt of material beyond Pluto. The orbit of a normal planet or asteroid around the Sun is an ellipse, but many comets seem to have a near parabolic orbit, shaped like an open-ended ellipse. No comet has been seen with a hyperbolic orbit, which would show that it came from outside the solar system. This suggests that comets fall in from a far distant part of the solar system. This remote region is called the Oort cloud.

The Kuiper belt (right) extends out from just beyond the orbit of Neptune to about three times as far away as Neptune in the same plane as the planets. It is made of icy bodies, some hundreds of miles across, which would make spectacular comets if they came close to the Sun. The belt is probably, in fact, the source of short-period comets, which generally orbit in the plane of the solar system. The first members of the belt were discovered in 1992 and 1993. By comparison with the surrounding Oort cloud, the Kuiper belt is tiny.

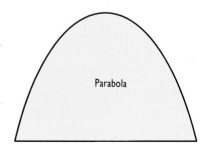

There are three shapes of orbit around the Sun which tell where a body comes from. A body tracing an ellipse is in a relatively close orbit around the Sun; a body in an almost parabolic orbit has come from the outer reaches of the solar system; a body following a hyperbola would probably come from outside the solar system.

Mathematicians have long known that all three shapes can be cut out of a cone (below), and the angle of the cut to its vertical axis determines the shape of the curve.

Hyperbolic path

A golf ball that swings around the lip of the hole and out again moves in a hyperbola, the path of a body that has enough energy to escape from the gravitational pull of the Sun. Any object with a hyperbolic orbit would be a visitor from outside the solar system; none have yet been seen.

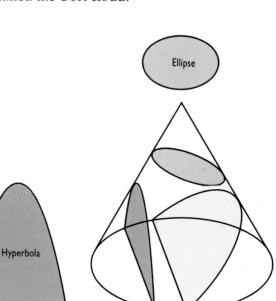

Ellipse

Hyperbola

Parabola

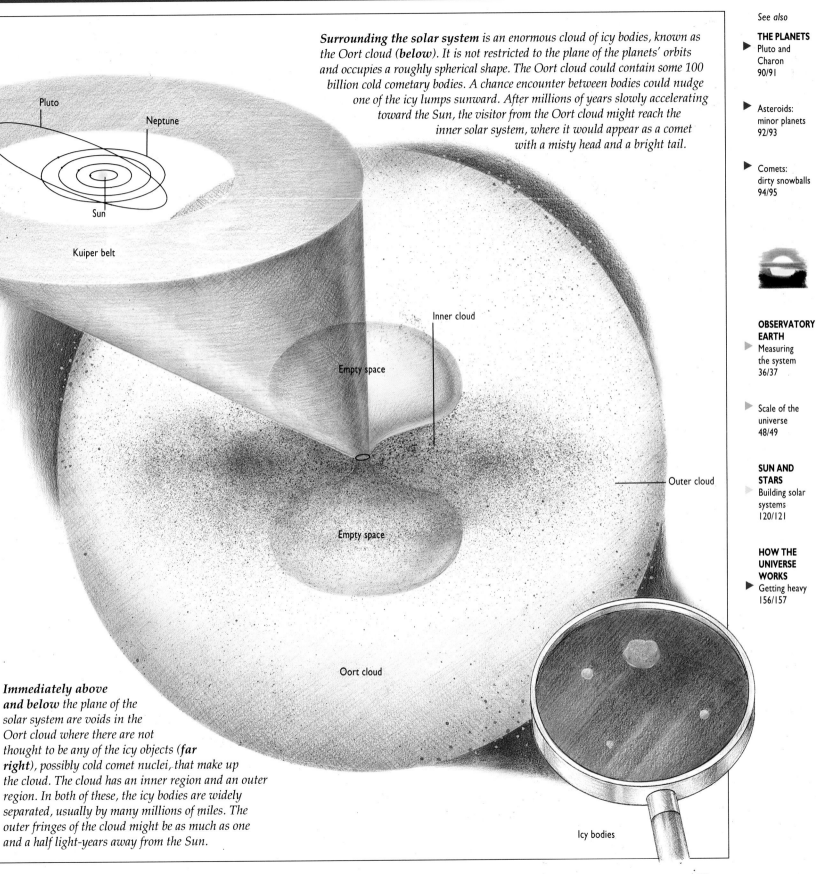

Pluto

Neptune

Sun

Kuiper belt

Inner cloud

Empty space

Empty space

Outer cloud

Oort cloud

Icy bodies

Surrounding the solar system is an enormous cloud of icy bodies, known as the Oort cloud (**below**). It is not restricted to the plane of the planets' orbits and occupies a roughly spherical shape. The Oort cloud could contain some 100 billion cold cometary bodies. A chance encounter between bodies could nudge one of the icy lumps sunward. After millions of years slowly accelerating toward the Sun, the visitor from the Oort cloud might reach the inner solar system, where it would appear as a comet with a misty head and a bright tail.

Immediately above and below the plane of the solar system are voids in the Oort cloud where there are not thought to be any of the icy objects (**far right**), possibly cold comet nuclei, that make up the cloud. The cloud has an inner region and an outer region. In both of these, the icy bodies are widely separated, usually by many millions of miles. The outer fringes of the cloud might be as much as one and a half light-years away from the Sun.

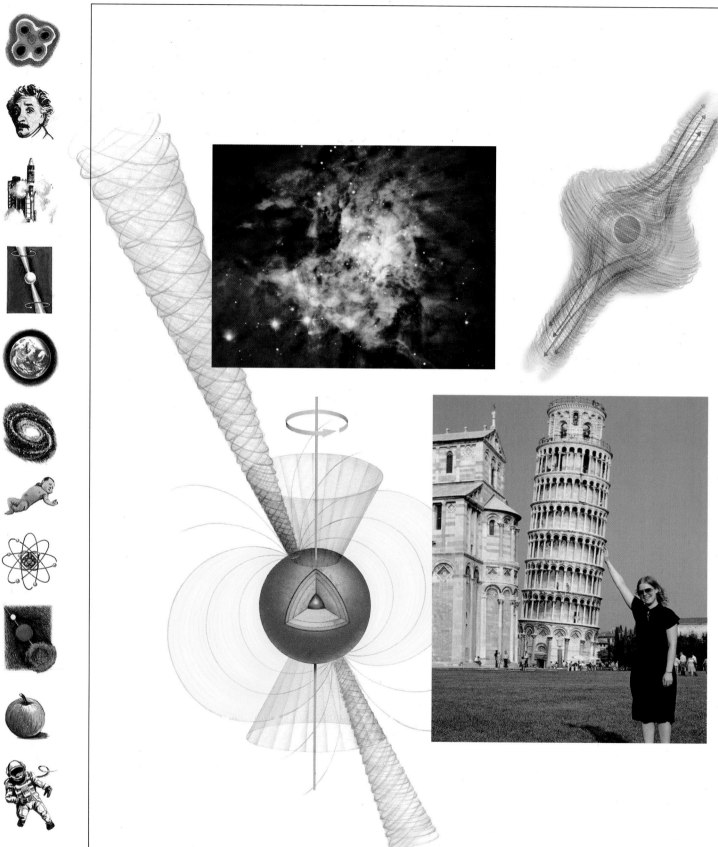

Sun and Stars

O ur sun is so *everyday, reassuring, and familiar* – shining steadily in the sky and providing us with light and heat – that it is almost unimaginable that within it, the energy from many millions of nuclear bombs is being generated each second. Deep in the core of our own local star, hydrogen atoms fuse together to make another element, helium. The energy released in the process warms the Earth today, as it has for billions of years and will for billions more.

The myriad points of light in the night sky are also suns and, like our sun, the stars have formed out of gas clouds and will eventually burn out and die. Between their birth pangs and death throes, stars obey a simple rule: the bigger they are, the brighter they shine. A really big star explodes cataclysmically at the end of its life and can leave behind a black hole, into which light itself can disappear forever.

Left (clockwise from top): multiple stars; a star is born from gas; juxtaposition can deceive when it comes to judging distance on Earth, and in the sky; a neutron star beaming radio pulses.
This page (top): inside the Sun where matter is destroyed to power a star; (left) the view from a black hole – looking out from the inside.

A regular star

Our very own star, the Sun, is stable, long-lived, and life-giving, but it is also a raging nuclear inferno.

If you lived on a planet orbiting Proxima Centauri, our nearest stellar neighbor at 4.3 light-years away, you would see our sun as a point of light, just like countless other stars. Its brightness and color would tell you that it was an ordinary star in the stable, middle period of its life. It would be classified as a dwarf star, rather than as a giant like Betelgeuse in the constellation of Orion, whose diameter is 500 times larger. Certainly, there would be nothing about it that would tell you it was in any way special.

In fact, in astronomical terms, the Sun is not particularly remarkable, but it means a great deal to Earth dwellers – without its heat and light we would perish. The Sun is the source of all life on Earth: its heat warms the planet, and its light is used by green plants in photo-synthesis, the start of the food chain.

Never observe the Sun directly. Even looking at it with the naked eye can cause pain and damage. Observing it through a telescope or binoculars will cause irreversible damage to the retina, the back of the eye, which will impair vision and probably cause partial blindness. Do not rely on filters sold for use with telescopes to protect you – they might fall off or crack.

*The only safe way to observe the Sun is to use a telescope to project an image of the Sun onto a sheet of white paper (**right**). The paper should be shielded from direct sunlight so the focused image of the Sun stands out in bright contrast. On the projection, any sunspots should be visible. From their shifting positions day by day, the rotation of the surface of the Sun can be followed.*

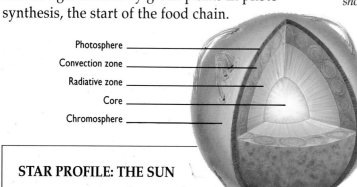

Photosphere
Convection zone
Radiative zone
Core
Chromosphere

Prominence

STAR PROFILE: THE SUN

	miles	km
Distance from Earth	92.96 million	149.6 million
Diameter	0.864 million	1.39 million
Age	4.5 billion years	
Mass	1.99×10^{27} tons	
Mean density	0.254 of Earth's	
Luminosity	3.9×10^{27} kW	
Surface temperature	5,800K	
Absolute magnitude	4.83	
Spectral type	G2	

00:00 60

00:00 30

00:00 00

Meteors and protoplanets begin to condense

Fusion in core approaches steady equilibrium

Fusion begins in core, star ignites

In addition to heat and light, however, the Sun gives off the full range of electromagnetic radiation, not all of which is life-giving. It is a source of deadly gamma rays, X-rays, and ultraviolet – as well as infrared and radio waves – but the Earth's magnetic field and atmosphere protect us from this dangerous, yet nurturing, star.

The Sun is not one of the first generation of stars formed after the Big Bang. Like countless others, it condensed from gas that was already laced with heavy elements. These elements were forged in the nuclear furnace of earlier, large stars and were blasted into space when these stars became supernovae. This means that there were heavy elements in the gas and dust from which the Sun and the planets condensed. Without these elements, our Earth, and all life on it, could not exist.

Planets solidify

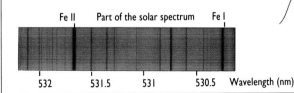

Protoplanets collect material

If the Sun's lifespan were a minute, starting from when the star ignited and fusion began in its core, all the most interesting developments would happen in just a few seconds at each end of that time period. After an initial erratic period lasting only about one-third of a second, the Sun would settle down. After three seconds, the planets would stabilize. On the clock, 27.6 seconds would represent now – when the Sun has been more or less stable for the last 4 billion years and is well into its middle age.

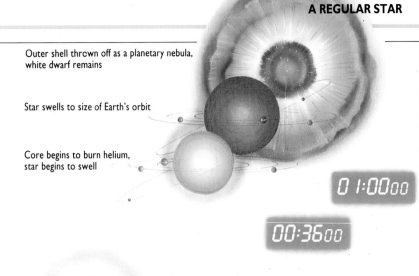

Outer shell thrown off as a planetary nebula, white dwarf remains

Star swells to size of Earth's orbit

Core begins to burn helium, star begins to swell

`01:00`oo

`00:36`oo

Now

`00:27`6o

Star burns steadily, core has burned up 65% of its hydrogen fuel

`00:03`oo

The Sun will continue much the same as it is now for the next 5 billion years. Slowly, as the hydrogen at its core gets used up, it will grow red and swell up until it enlarges to have a diameter about that of the width of the Earth's orbit. Some time after that, the Sun will blow off its outer layers to leave behind an intensely hot, bright, small white dwarf, and its minute will be over.

LINES IN THE SUNLIGHT

Fraunhofer made an accurate map of the dark lines (Fraunhofer lines) in the Sun's spectrum. Robert Bunsen and Gustav Kirchhoff later deduced that these are "fingerprints" of elements. Spectral lines reveal the elements that are present in an incandescent body like the Sun. Iron (Fe), for instance, always shows lines at the same wavelengths (**below**).

In 1868, solar spectrum lines were seen that did not match those of any element then known on Earth. This new element was named helium, after *helios*, Greek for the Sun.

The German physicist Joseph *von Fraunhofer (1787–1826), who studied optics and founded an optical institute in 1807, began studying the dark lines in the Sun's spectrum in 1814. He died 12 years later, at the age of 39, but others continued his pioneering work.*

Fe II   Part of the solar spectrum   Fe I

532   531.5   531   530.5   Wavelength (nm)

The face of the Sun

Our sun has a variety of different features from dark spots to vast gleaming arches of incandescent gas.

Although the energy of the Sun comes from deep within it, the light we see comes from an outer shell, called the photosphere, just 185 miles (300 km) thick. It is only through this outer layer that light, heat, and other types of radiation can escape. The photosphere glows because it is heated to nearly 5,800K by the nuclear reactions in the core.

The photosphere is not uniformly bright, and in it darker blotches can often be observed that are 1,500K or so cooler. These blotches are called sunspots, and they occur in pairs. The movement of sunspots across the Sun's disk shows that it is rotating. The rate is not uniform, however, and varies from 26 days near the poles to 37 days at the equator.

High up in the photosphere, there are also hotter, brighter spots, or faculae, which occur in large numbers in the vicinity of new sunspots, often above where a sunspot is about to appear. Other bright spots, called flocculi, burst into solar flares, which give off so much radiation that their effects can be felt as magnetic storms on Earth. Huge looping structures, or prominences, usually hang quiescent in the magnetic fields

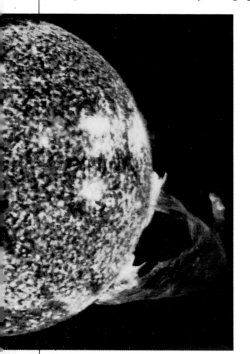

Solar prominences look like arches of flame (**left**). They are not flame, but incandescent gas which follows the lines of magnetic force that connect a pair of sunspots (**right**).

Sunspots last for anything from a couple of hours to a few months. The appearance of a pair of "pores," separate in longitude but roughly the same latitude, usually signals the birth of a sunspot.

Each sunspot pair has a leader and a follower, of different magnetic polarity. Reaching maximum size in 9 to 10 days, they may be as small as 600 miles (1,000 km) across, or large enough to swallow the Earth.

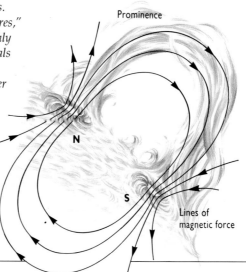

Prominence

N

S

Lines of magnetic force

The halo visible around the Sun when the blinding disk of the photosphere is eclipsed by the Moon is the corona (**above**), the Sun's outer atmosphere. The corona is 1 million times thinner than Earth's atmosphere, but the particles in it are energetic, giving the corona a temperature of about 1 million K. The inner corona spreads out for some 930,000 miles (1.5 million km), and there are also large jets with bulbous bases which spread out about 1.9 million miles (3 million km), following the lines of the Sun's magnetic field.

near sunspots, but they, like flares, can suddenly surge up at 600 miles/s (1,000 km/s) hundreds of thousands of miles into space.

Above the photosphere is the chromosphere. This is around 3,000 miles (5,000 km) thick and at its base has a temperature of approximately 4,000K. Throughout the chromosphere are spicules, flamelike structures 3,000–6,000 miles (5,000–10,000 km) long, which burst from the Sun's surface at a speed of about 16 miles/s (25 km/s). Also, dark red lines or filaments appear in the chromosphere above sunspots.

The Sun goes through a number of cycles in which it either changes size or shape, or vibrates. There seems to be an 80-year cycle during which its overall diameter contracts and expands by 0.02 percent. It also wobbles at different rates, rather like jelly: the polar and equatorial diameters bulge alternately by a few miles with a period of 52 minutes, and there is another slower oscillation with a period of 2 hours 40 minutes. Finally, there is a surface ripple with a 5-minute period.

*The outer layer of the Sun – the photosphere – looks like a simmering pan of soup (**above**). Soup, warmed by heat from below, expands and rises in bubblelike shapes called convection cells. In the Sun, convection cells give its surface a grainy appearance.*

At the center of each cell, which can measure 600–1,200 miles (1,000–2,000 km) across, the photosphere is rising at around ½ mile/s (1 km/s). Individual convection cells last for about 10 minutes before dispersing.

THE PATTERN OF SPOTS

Sunspot activity follows an 11-year cycle from one time when the number of sunspots is at a maximum to the next. The magnetic polarity of the sunspots changes in each succeeding cycle, so in one cycle the leading spot in each sunspot pair will have a north polarity and the trailing a south polarity; in the next cycle, the opposite will be the case. Thus there is a 22-year cycle superimposed on the 11-year cycle.

The sunspot cycle slightly affects the power output of the Sun, which increases when there are more spots. This in turn affects weather on Earth, with the result that an 11-year cycle can be seen in the growth rings in trees and in the patterns of thawing and freezing in glaciers.

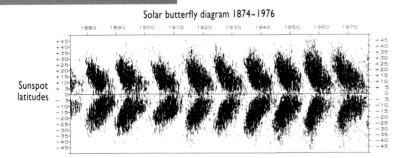

Solar butterfly diagram 1874–1976

*If the latitudes at which sunspots appear are plotted on a graph against time, a pattern like butterfly wings emerges (**above**). This happens because each cycle begins with the appearance of spots at latitudes 25–30° north or south of the Sun's equator. As the cycle progresses, succeeding spots form closer to the equator, but are uncommon on the equator itself.*

The nuclear powerhouse

Our sun is a ferocious ferment of energy that owes its power to reactions between minuscule particles deep in its interior.

In the center of the Sun, huge temperatures and pressures push protons – the nuclei of hydrogen atoms – so close together that some of them fuse. Through a series of reactions, four hydrogen nuclei are joined to produce one nucleus of helium. But one helium nucleus is only 3.97 times the mass of four protons, so during this reaction 0.0075 grams per gram of fused hydrogen are lost altogether, or annihilated. Matter that is annihilated turns to energy, and the amount of energy released is given by Einstein's equation $E=mc^2$ – in which energy equals mass annihilated times the speed of light squared. So for each gram of helium made, the 0.0075 grams of matter lost produce 6.75×10^{11} joules – or enough energy to run a 1 KW electric heater for about 20 years.

More than 600 million tons of hydrogen are being converted into helium each second in the Sun's core – its innermost 30 percent. In the process, more than four million tons of matter are converted into energy. But this huge loss per second is trivial compared to the total mass of the Sun. It takes, in fact, about 11 billion years to convert the hydrogen in a Sun-sized star's core to helium. The Sun is now about halfway through its supply.

The hydrogen bomb uses reactions like those in the Sun. A chemical explosion flings together uranium to set off an atomic explosion, which briefly concentrates enough pressure and heat on some heavy hydrogen to set off a fusion reaction and massive blast.

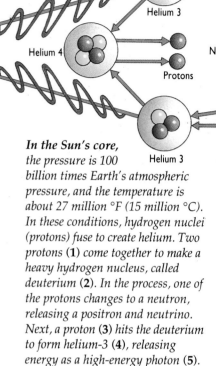

In the Sun's core, the pressure is 100 billion times Earth's atmospheric pressure, and the temperature is about 27 million °F (15 million °C). In these conditions, hydrogen nuclei (protons) fuse to create helium. Two protons (1) come together to make a heavy hydrogen nucleus, called deuterium (2). In the process, one of the protons changes to a neutron, releasing a positron and neutrino. Next, a proton (3) hits the deuterium to form helium-3 (4), releasing energy as a high-energy photon (5). Finally, two helium-3 nuclei come together to make a stable helium-4 nucleus, and two protons are released.

POWER LIKE THE SUN

Scientists are trying to tame and harness the fusion reactions that power the Sun. The hydrogen fuel is abundant in sea water, and the by-product is harmless helium. But scientists cannot match the pressure in the Sun's core, so they have to raise the temperatures to up to 300 million K. At the Joint European Torus (JET) project at Culham, England, hydrogen plasma is contained in a magnetic field inside a doughnut-shaped vacuum chamber. During its first run, in 1991, it made 2 MW of electricity.

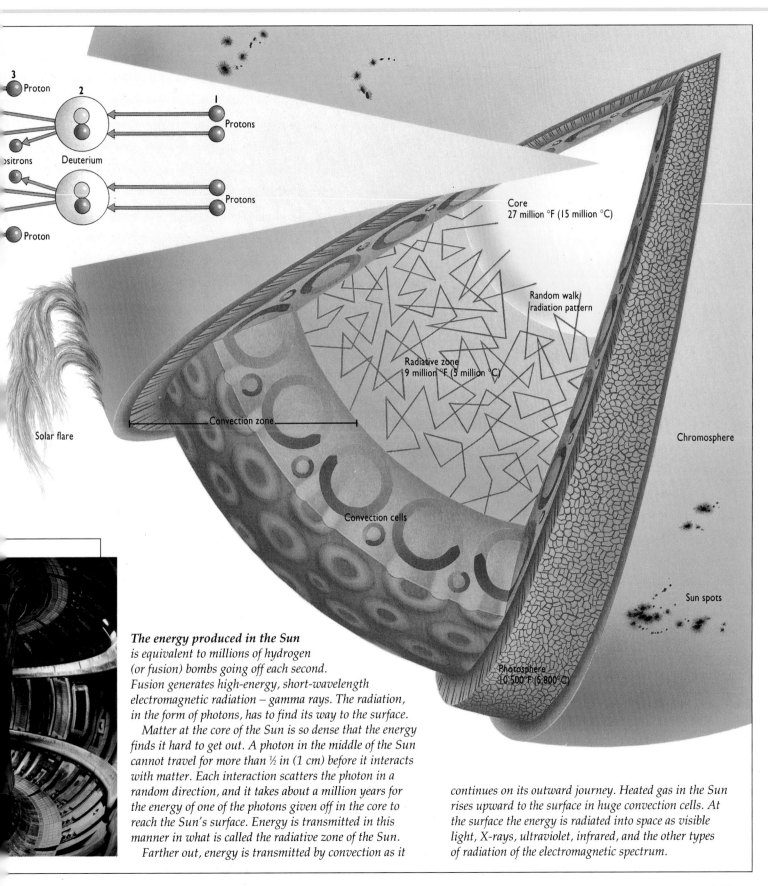

3 Proton

2

1

Protons

Deuterium

Positrons

Protons

Proton

Core
27 million °F (15 million °C)

Random walk
radiation pattern

Radiative zone
9 million °F (5 million °C)

Convection zone

Solar flare

Convection cells

Chromosphere

Sun spots

Photosphere
10,500 °F (5,800 °C)

The energy produced in the Sun
is equivalent to millions of hydrogen
(or fusion) bombs going off each second.
Fusion generates high-energy, short-wavelength
electromagnetic radiation – gamma rays. The radiation,
in the form of photons, has to find its way to the surface.

Matter at the core of the Sun is so dense that the energy
finds it hard to get out. A photon in the middle of the Sun
cannot travel for more than ½ in (1 cm) before it interacts
with matter. Each interaction scatters the photon in a
random direction, and it takes about a million years for
the energy of one of the photons given off in the core to
reach the Sun's surface. Energy is transmitted in this
manner in what is called the radiative zone of the Sun.

Farther out, energy is transmitted by convection as it
continues on its outward journey. Heated gas in the Sun
rises upward to the surface in huge convection cells. At
the surface the energy is radiated into space as visible
light, X-rays, ultraviolet, infrared, and the other types
of radiation of the electromagnetic spectrum.

Center of the system

The aurorae of Earth's polar regions are one of the most spectacular manifestations of the many ways in which the Sun influences its captive retinue of bodies.

Not only does the Sun bind the solar system together with its gravitational force, it also bathes it with energy and showers it with particles. In fact, the Sun flings 1 million tons of matter out into space every second, creating the solar wind.

Matter in the wind is made mostly of three types of particles: protons, or hydrogen ions (atoms that have had their electrons stripped away); alpha particles, or helium ions; and electrons. All three are electrically charged.

The Sun also emits radiation from the full range of the electromagnetic spectrum, including short-wavelength, penetrating gamma rays and X-rays, ultraviolet, visible light, and the longer wavelengths, such as infrared, microwave, and radio waves. The total amount of energy of all wavelengths we would receive on Earth if the atmosphere was not there is the solar constant, which is 1,144 watts/sq yd (1,368 watts/m^2). If all the energy received by an area the size of a football field could be converted into electricity, it would power a small town.

Small variations occur in the constant, and it is thought that the Sun's luminosity varies by ½–1 percent over time. These variations could account for the ice ages. A decrease of just 2½ percent in the output of the Sun could create a permanent ice age. Ice would cover so much of the Earth and might reflect back so much sunlight that even if the solar constant returned to its present level, the Earth might never thaw.

The solar wind is a continuation of the corona, the Sun's outer atmosphere, which is blasted off the surface of the Sun in huge jets. Because the Sun spins, the solar wind stream off in huge spirals of particles, like the water from a garden sprinkler (above). The particles leave the Sun at up to 3,000 miles/s (5,000 km/s), but they are slowed down by the Sun's gravity and magnetism so that by the time they reach Earth, after four or five days, they are traveling at up to 435 miles/s (700 km/s). The temperature of the particles is extremely high at 50,000–500,000K.

Like winds on Earth, the solar wind blows in gusts, especially when solar flares erupt. These send out extra particles which can create magnetic storms that may affect radio transmission and can cause power failures.

Without the protection of our atmosphere, a space-walking astronaut has to wear a suit designed to guard against the solar wind and radiation. But the metal-lined suit offers little protection against protons and alpha and beta particles; neither does it do much to stop gamma rays – which are like high energy X-rays. To reduce risk, space walks are not made after solar flares, when levels are high. Even when there are no flares, the radiation in spacecraft is 20 times the approved "safe" level on Earth.

The northern lights, or *aurora borealis, flicker and shimmer in the northern night sky with colorful glowing streamers and curtains of light (***left***). The equivalent in the southern hemisphere is the aurora australis, or southern lights.*

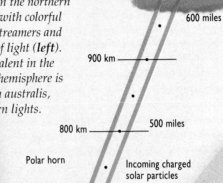

Aurora caused by ionized nitrogen emitting violet/blue light

1,000 km

600 miles

900 km

800 km — 500 miles

Polar horn

Incoming charged solar particles

700 km — 400 miles

600 km

500 km — 300 miles

400 km — 200 miles

Aurora caused by oxygen atoms emitting red light

Magnetic force lines

300 km

Magnetic force lines

200 km — 100 miles

The color of the aurorae depends on the type of gas that is glowing and its height in the Earth's atmosphere. Oxygen, for instance, can give green or red light depending on the altitude.

100 km

Aurora caused by oxygen atoms emitting green light

Aurora caused by molecular nitrogen emitting red light

Earth

THE GLOWING SKY AND THE WIND

The Earth's magnetic field steers ionized (electrically charged) particles of the solar wind around the planet, collecting them in the Van Allen radiation belts. Fortunately, not many of these potentially harmful particles can penetrate the Earth's atmosphere.

The Earth's magnetic force lines are curved, and at the poles they come down to the surface. This allows some solar wind particles to reach the upper parts of the atmosphere. When an ionized particle moving at speed hits a gas molecule it, in turn, ionizes the molecule (that is, knocks off an electron or two). This can also make gas molecules glow, causing the aurorae. When there is much activity on the Sun – for instance, when there are flares associated with sunspots – more particles are given off, and the aurorae can be really spectacular. Other planets with magnetic fields, such as Jupiter and Saturn, also have aurorae.

Stars in motion

The Sun and its fellow travelers orbiting the Galaxy move through space like a scattered swarm of bees.

Gaps between stars are thousands of times bigger than the distance between the Sun and Pluto, the most distant planet of the solar system. It is so far to the next star (Proxima Centauri) that light moving at 186,300 miles/s (300,000 km/s) takes 4.3 years to reach Earth. Near the Sun, the stars seem to have gaps of about 3–4 light-years between them, and Proxima Centauri is just one of the couple of dozen or so relatively unremarkable stars within about 10 light-years of our sun.

The center of our galaxy contains about 10 billion stars, and the Sun and its neighbors are all moving through space as they orbit around it. In fact, relative to the Galaxy's center, the Sun is moving at about 513,000 mph (825,000 km/h). While stars are going in the same general direction, they are not in the same orbit and so move relative to each other. Each star in the sky thus has what is known as a proper motion. This is the gradual change in its position against the stellar background caused by the difference between its orbit around the center of the Galaxy and that of the Sun.

Although the positions of the stars do change, the movement of most of them is almost imperceptibly slow. Changes can be measured only on photographs taken many years apart through powerful telescopes. The fastest moving star we can see is nearby Barnard's star, a 9th-magnitude red dwarf in the constellation of Ophiuchus. This is the second closest star to our sun, at 6 light-years away, and it moves across the sky by 10.3 arc seconds each year (one degree every 350 years). Most proper motions are many times slower than this. However, over long periods the patterns of stars that make up the constellations will change because, by and large, constellations are chance juxtapositions of bright stars in our line of sight from Earth, not physically linked groups that move together.

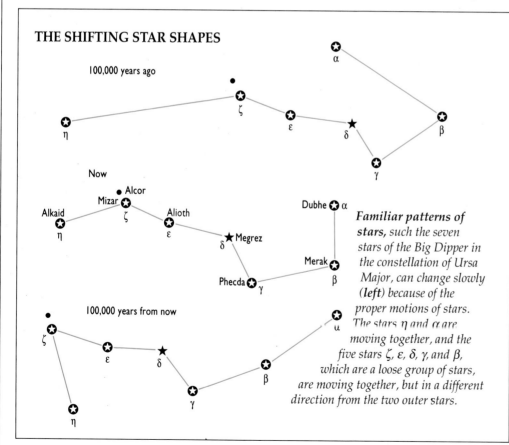

THE SHIFTING STAR SHAPES

100,000 years ago

Now

100,000 years from now

Familiar patterns of stars, such the seven stars of the Big Dipper in the constellation of Ursa Major, can change slowly (left) because of the proper motions of stars. The stars η and α are moving together, and the five stars ζ, ε, δ, γ, and β, which are a loose group of stars, are moving together, but in a different direction from the two outer stars.

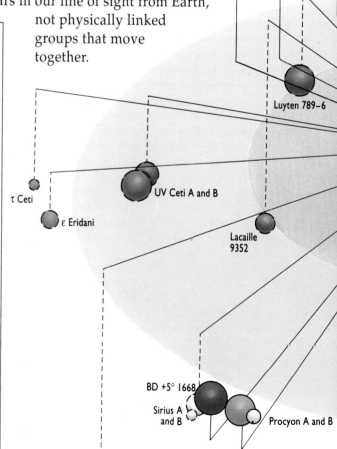

Kruger 60 A and B

Groombridge 34 A and B

Ross 248

Luyten 789–6

τ Ceti

ε Eridani

UV Ceti A and B

Lacaille 9352

BD +5° 1668

Sirius A and B

Procyon A and B

Kapteyn's star

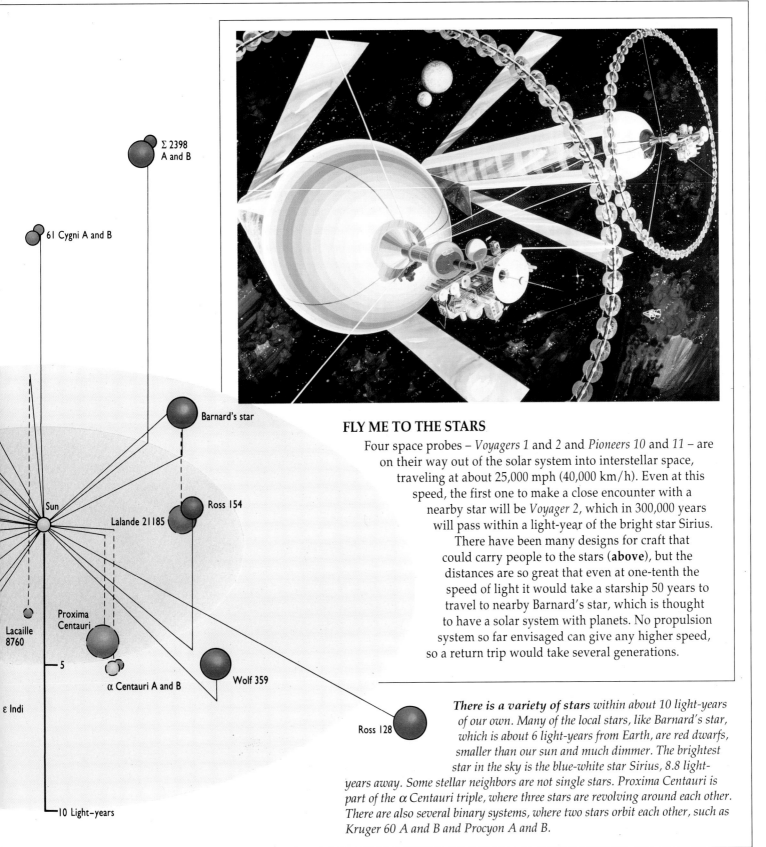

FLY ME TO THE STARS

Four space probes – *Voyagers 1* and *2* and *Pioneers 10* and *11* – are
on their way out of the solar system into interstellar space,
traveling at about 25,000 mph (40,000 km/h). Even at this
speed, the first one to make a close encounter with a
nearby star will be *Voyager 2*, which in 300,000 years
will pass within a light-year of the bright star Sirius.
There have been many designs for craft that
could carry people to the stars (**above**), but the
distances are so great that even at one-tenth the
speed of light it would take a starship 50 years to
travel to nearby Barnard's star, which is thought
to have a solar system with planets. No propulsion
system so far envisaged can give any higher speed,
so a return trip would take several generations.

*There is a variety of stars within about 10 light-years
of our own. Many of the local stars, like Barnard's star,
which is about 6 light-years from Earth, are red dwarfs,
smaller than our sun and much dimmer. The brightest
star in the sky is the blue-white star Sirius, 8.8 light-
years away. Some stellar neighbors are not single stars. Proxima Centauri is
part of the α Centauri triple, where three stars are revolving around each other.
There are also several binary systems, where two stars orbit each other, such as
Kruger 60 A and B and Procyon A and B.*

Σ 2398
A and B

61 Cygni A and B

Barnard's star

Sun

Ross 154

Lalande 21185

Proxima
Centauri

Lacaille
8760

5

α Centauri A and B

Wolf 359

ε Indi

Ross 128

10 Light-years

Star bright

A star's real brightness is not the same as the brightness it appears to have in the night sky.

Even a glance at the night sky shows that the stars are not all equally bright. In a very early star catalog, compiled in 150 B.C., Greek astronomer Hipparchus grouped the stars in six "sizes" – or magnitudes – in decreasing order of brightness. The brightest stars were magnitude 1, the next brightest were magnitude 2, and so on down to magnitude 6 for the faintest objects that could be seen with the naked eye. Hipparchus's scale is still in use today, but has been extended and refined. The difference in brightness over 5 magnitudes is now set at 100 times, which means that a star one magnitude less than another is, in fact, 2.512 times brighter.

Magnitude 6 is still the limit of visibility for "naked eye" objects. With binoculars, however, it is possible to see to magnitude 9, and with an amateur's 3-inch (75-mm) telescope to magnitude 11. The 200-inch (5.08-m) telescope at the Mount Palomar Observatory makes it possible to see stars of magnitude 20.6 and to photograph stars of magnitude 23.5. The Hubble Space Telescope can detect stars of magnitude 29.

If a person is in the same line of sight as a large distant object, the two can appear to be almost the same size, as in this photograph of someone "holding up" the Leaning Tower of Pisa. The true size of the person relative to the tower becomes clear only if tower and person are at an equal distance from the viewer.

A similar principle applies to star brightnesses. A dim star near Earth can seem as bright as a luminous, distant one. Relative brightnesses can be assessed only when the stars are at the same distance.

Human imagination conjured a hunter, with his belt, sword, club and shield, out of the familiar pattern of stars that makes up the constellation of Orion (right). Although a constellation such as Orion seems to have a clear pattern when viewed from Earth, the stars are all at different distances from us. In fact, if the stars in Orion were viewed from elsewhere in our galaxy, the constellation would lose its distinctive shape completely.

Saip

Orion

In a projection that shows them at their true distances from Earth, it is clear that the seven main stars of Orion are not physically close to each other at all (right), but are separated by huge distances. The farthest star in the pattern, Saiph, is actually more than four times as far away as the star closest to the Earth, Bellatrix.

The Great Nebula in Orion, M42, which hangs from the belt, is just visible to the naked eye as a dim patch of pearly light on a clear night. This region, where stars are being formed, is about 1,000 light-years from Earth, but if it were just 30 or so light-years distant, it would occupy a large section of the sky.

Vertical lines (left) link the positions in space of the stars, marked by their Greek letters, to the plane at the top of the diagram where the stars are marked by their proper names. To read off the distances of the stars from the Earth, trace a horizontal line from where the star is named to the scale in light-years.

Light-years
2,100
1,950
1,800
1,650
1,500
1,350
1,200
1,050
900
750
600
450
300
150
0

Alnilam
Mintaka
Alnitak
Great Nebula
Rigel
Betelgeuse
Bellatrix

α
γ
ζ ε δ
Great Nebula
β
κ

Betelgeuse α

Bellatrix γ

Mintaka δ

Alnilam ε

Alnitak ζ

Saiph κ

Rigel β

At 32.6 light-years (10 parsecs) away, all seven of Orion's main stars would be much brighter than we see them (right) because they are very luminous stars with high absolute magnitudes. At this distance Saiph, apparently the dimmest of the seven main stars with an apparent magnitude of 2.06, would have a magnitude of –6.9, making it the second brightest after Rigel, which has an absolute magnitude of –7.1.

This is 23 magnitudes less than the faintest object visible with the naked eye and is thus 2.512^{23} – about 1.6 billion – times less bright.

The scale has also been extended in the other direction. The bright star Vega in the constellation Lyra, for example, is close to magnitude 0. Even brighter objects have negative magnitudes, so Venus is about magnitude –4, the full Moon is about –12 and the Sun is –26.5.

Measuring the brightness of an object in the night sky only gives its apparent magnitude, however – that is, how bright it appears from Earth. To determine how bright an object actually is, astronomers work out its absolute magnitude. This is the brightness it would have at a distance from the Earth of 32.6 light-years (10 parsecs).

At that distance, the luminous, far-off star Betelgeuse, with an apparent magnitude of 0.8, would have an absolute magnitude of about –6. The Sun, by contrast, would have a magnitude of just 4.83, making it visible to the naked eye, but not particularly conspicuous.

Colors of the stars

A star's hue not only reveals its surface temperature, but can also indicate its mass and life expectancy.

With the naked eye, it is obvious that stars are not all the same color. Rigel (Beta Orionis) is blue, Sirius (Alpha Canis Majoris) is white, our sun is yellowish, Aldebaran (Alpha Tauri) is orange, and Betelgeuse (Alpha Orionis) is red. Stars have different colors because they have different surface temperatures. Blue stars are hotter than white stars which are hotter than yellow stars. These in turn are hotter than orange stars, which are hotter than red stars.

Each color of star has its own type of spectrum and is classified using the letters O, B, A, F, G, K, M, which define its spectral class. Blue O-class stars have surface temperatures of over 25,000K. B-class stars have surface temperatures of 25,000 to 11,000K and are bluish-white, and so on down to red M-class stars which have surface temperatures below 3,500K.

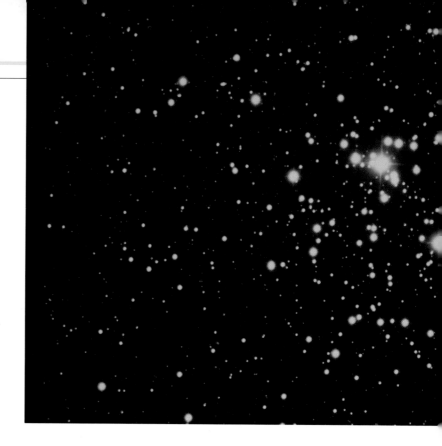

Each class is further divided into 10 numbered sub-classes, in which 0 is the hottest and 9 the coolest. Our sun, with a surface temperature of around 5,800K, is a yellow G2 star.

Early in the 20th century, the Danish astronomer Ejnar Hertzsprung and the American Henry Russell plotted the positions of stars on a graph of absolute magnitude (a star's luminosity relative to others) against surface temperature. They found that stars group in some areas and are absent from other areas of the graph, known as the Hertzsprung–Russell (H–R) diagram.

Most stars fall into the "main sequence," a grouping which wends its way diagonally across the diagram, from dim red stars at the bottom right to bright blue stars at the top left. The Sun falls roughly in the middle of the main sequence. A star's position on the sequence depends on its mass, with the least massive stars at the bottom right of the diagram and the most massive at the top left, because more massive stars burn hotter. Below the main sequence is a line of dim white stars, white dwarfs; above is a group of variable stars and luminous red stars, known as giants. Across the top of the diagram are a line of extremely luminous stars, the supergiants.

The color of a glowing object tells us about its temperature. In a foundry (left), when a metal is heated, it first glows dull red, then bright red, then orange, yellow, and white as it becomes increasingly hotter.

On the main sequence of the H–R diagram, the bright stars are blue and the dim red. A star's color (and thus temperature and spectrum) depends on its mass.

B-type star

A star of class B (left) on the main sequence is one of the most massive. It burns fast, bright, and blue with a surface temperature of between 11,000 and 25,000K. In its spectrum (left), the hotter, bluer colors predominate. Blue stars have few spectral absorption lines. The biggest blue stars on the sequence are about 20 times the diameter of our sun and 50 times its mass, but millions of times brighter.

A G-class star (right), like our sun, is yellow and can have a surface temperature between 4,700 and 6,000K. The middle colors of visible light dominate the spectrum. Stars of this size burn steadily for about 10 billion years, gradually swelling up to become red giants, which are more luminous because of their greater diameter. In time they become white dwarfs.

G-type star

An M-class star (left) is smaller, cooler, and redder than the Sun with a temperature of about 3,500K. Red wavelengths dominate, and there are many absorption lines visible in its spectrum. M-stars are long-lived, shining for 100 billion years or more.

M-type star

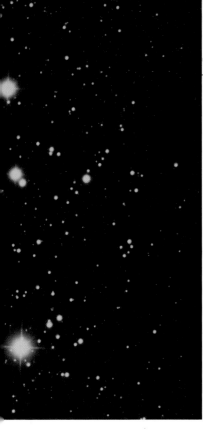

The Jewel Box cluster in the constellation Crux (above) was described by the British astronomer John Herschel (1792–1871), son of the famous William Herschel (1738–1822), as a "casket of precious stones." It shows vividly the different colors that stars can have, the blues, whites, and pinkish-reds of the stars matching their surface temperatures – bluer is hotter.

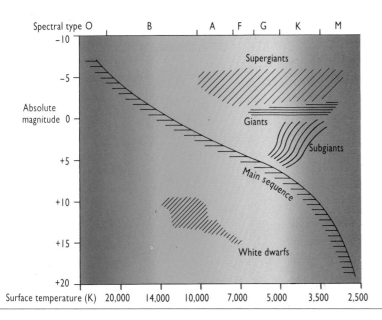

The Hertzsprung–Russell diagram (left) is a snapshot of the galaxy as we see it today; some stars on it are new and some old, but it can tell us a lot about how stars evolve. New stars appear just above the main sequence. As they settle to steady hydrogen fusion, they stabilize on the sequence, with less massive stars burning cool, red, and dim and more massive ones burning hot, blue, and bright. The bluer a star on the sequence, the shorter its life.

Older stars move off the main sequence when their fusion reactions change, becoming luminous giants or supergiants. Stars in this stage are unstable, and their output changes make them variables. The more massive a star is at "birth," the brighter it is at the giant stage. As a star reaches the end of its giant phase, it gets redder and moves to the right of the diagram. Supergiants eventually explode and disappear off the diagram. Stars the size of our sun end up as dim white dwarfs below the main sequence after they have ceased fusion in their cores.

See also

SUN AND STARS
▶ A regular star 102/103

▶ The nuclear powerhouse 106/107

▶ Stars in motion 110/111

▶ Star bright 112/113

▶ Double stars 116/117

▶ Variable stars 118/119

▶ Bang or whimper? 122/123

OBSERVATORY EARTH
▶ Light-years 38/39

▶ Deep space 40/41

▶ Putting light to work 42/43

NEBULAE AND GALAXIES
▶ Star clusters 136/137

HOW THE UNIVERSE WORKS
▶ Making waves 160/161

Double stars

Seen with the naked eye, some stars appear double, and most, it seems, are not alone.

One thing that is odd about our otherwise unremarkable sun is that it has no companion star. It is thought that more than 50 percent of all stars are, in fact, multiple systems made up of more than one sun. About three-fifths of these multiple systems are binaries (have two suns), three-tenths are triples, and one-tenth are made of four or more stars. Of the 12 stars closest to us, our sun appears to be the only single one.

There are several different types of double, or binary, stars, classified according to how close they are to each other and how they are seen from Earth. The four most interesting types are the optical double, the eclipsing binary, the astrometric binary, and the spectroscopic binary. Stars can also exist in complex groups with three and sometimes more members.

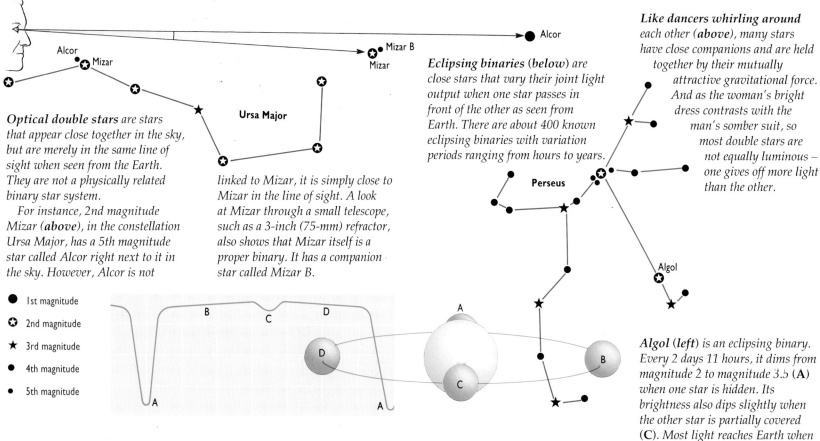

Optical double stars *are stars that appear close together in the sky, but are merely in the same line of sight when seen from the Earth. They are not a physically related binary star system.*

*For instance, 2nd magnitude Mizar (**above**), in the constellation Ursa Major, has a 5th magnitude star called Alcor right next to it in the sky. However, Alcor is not* linked to Mizar, it is simply close to Mizar in the line of sight. A look at Mizar through a small telescope, such as a 3-inch (75-mm) refractor, also shows that Mizar itself is a proper binary. It has a companion star called Mizar B.

Eclipsing binaries (**below**) *are close stars that vary their joint light output when one star passes in front of the other as seen from Earth. There are about 400 known eclipsing binaries with variation periods ranging from hours to years.*

Like dancers whirling around *each other (**above**), many stars have close companions and are held together by their mutually attractive gravitational force. And as the woman's bright dress contrasts with the man's somber suit, so most double stars are not equally luminous – one gives off more light than the other.*

- ● 1st magnitude
- ✪ 2nd magnitude
- ★ 3rd magnitude
- ● 4th magnitude
- • 5th magnitude

Algol (**left**) *is an eclipsing binary. Every 2 days 11 hours, it dims from magnitude 2 to magnitude 3.5 (**A**) when one star is hidden. Its brightness also dips slightly when the other star is partially covered (**C**). Most light reaches Earth when neither star is obscured (**B** and **D**).*

Path of single star

Path of star with faint or invisible companion

Sirius

Canis Major

Path of Sirius A   Path of Sirius B

Astrometric binaries are stars in which the presence of a companion can be detected by tracking the star's motion. A single star moves in a straight line, but a binary can wobble across the sky (**left**). This occurs because the stars of a binary both orbit the system's center of gravity, which moves in a straight line.

Sirius is such a binary: bright Sirius A and barely detectable Sirius B. Plot the motions of Sirius A (orbit shown in small circles) as it moves around the binary's center of gravity opposite Sirius B (orbit shown in large circles), and it traces a wave-shaped path across the sky.

In a spectroscopic binary like Capella (**below right**), two stars are so close to each other and so far distant from Earth that they cannot be separated visually even when observed through the most powerful telescopes. The only way that the presence of two stars can be revealed is by analysis of the spectrum (**right**), in which the spectrum of one star is superimposed on the other. Spectroscopic binaries can be detected only when the plane of the stars' orbit is seen more or less sideways from Earth, so that each star alternately moves away from and then toward the Earth as it orbits around the center of gravity of the binary. Because they are so close together, the speeds of the stars as they orbit are extremely high. The stars move so fast, in fact, that the lines of the spectrum of each are shifted first one way then the other, as seen from Earth, because of the Doppler effect.

At **1** and **3**, the stars are moving across the line of sight as seen from Earth and are thus moving neither away from nor toward the Earth, so the lines in each star's spectrum are not shifted, and the lines in their joint spectrum remain superimposed. At **2** and **4**, one star is moving fast toward Earth, so its spectral lines are shifted toward the blue end of the spectrum, while the other is moving fast away from Earth, so its spectral lines are shifted toward the red end of the spectrum. This means that the lines of the stars are shifted in opposite directions, and in the joint spectrum, this has the effect of separating the lines.

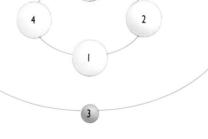

Capella

Auriga

The Trapezium, a multi-member star (**left**) in the Great Nebula in the constellation Orion, is a clutch of young, massive blue stars. The brightest stars of the system can be picked out in the middle of the photograph from amid the glowing gas and shining dust of this star-formation region.

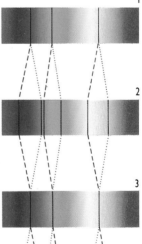

Variable stars

Not all stars shine steadily, so why do some change their brightness?

Many variable stars are those that are at either the beginnings or ends of their lives. At these times, the amount of light a star gives out can change dramatically as it settles down before beginning its stable, hydrogen-burning phase or, in an old star, as it destabilizes when its internal processes change because the hydrogen in its core becomes used up. Light output can change because a star expands and contracts, because it changes its surface temperature, or because layers of matter temporarily reduce its output.

Variables are classified in groups, usually named after a star that is typical of others of that type, or after the first star of that classification to be closely observed. An important group of variables is the Cepheids, named after Delta Cephei; they vary regularly over time and in their light output. The time taken for the variation – which ranges between 10 and 20 percent of output – is directly related to the brightness, or absolute luminosity, of the star.

Wolf-Rayet stars are luminous, erratic blue variables. Their absolute magnitudes are around –5 with surface temperatures of 30,000–50,000K. They are old stars that have lost their atmospheres, and their observable variations are the result of the violent processes that take place in the core of more stable stars, but are normally hidden by the atmosphere.

In one type of variable called R Coronae Borealis stars, the brightness suddenly decreases by up to 10 magnitudes, then slowly returns to normal. This is believed to be because solid carbon – soot – accumulates in the star's atmosphere. Energy builds up

Some variable stars have uneven surface temperatures and dark or bright spots. The variation in their light output has the same period as their rotation. As the dark region is carried around the star's surface by the rotation of the star, the light output temporarily reduces (*right*).

Our sun varies almost imperceptibly in this way because sunspots change its brightness by a tiny amount. Sunspots have a regular cycle: their numbers increase every 11 years or so. At these times, the Sun's output drops by a small amount, depending on how many spots it has – the more spots there are on its surface, the less the Sun's output.

Like a geyser that periodically shoots forth hot water (*above*), a regularly variable star of the pulsating type puts on a display at fixed intervals. Although the mechanisms of a geyser and a variable star are very different, they both release extra energy on a regular basis.

Old Faithful in Yellowstone National Park, Wyoming, is the world's most famous geyser, visited by many tourists annually.

Bright spot

1

2

Rotation direction

3

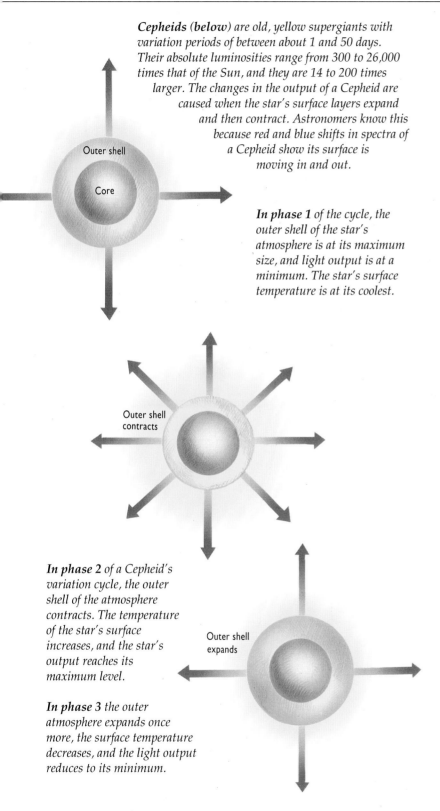

Cepheids (below) are old, yellow supergiants with variation periods of between about 1 and 50 days. Their absolute luminosities range from 300 to 26,000 times that of the Sun, and they are 14 to 200 times larger. The changes in the output of a Cepheid are caused when the star's surface layers expand and then contract. Astronomers know this because red and blue shifts in spectra of a Cepheid show its surface is moving in and out.

In phase 1 of the cycle, the outer shell of the star's atmosphere is at its maximum size, and light output is at a minimum. The star's surface temperature is at its coolest.

Core

Outer shell

Outer shell contracts

Outer shell expands

In phase 2 of a Cepheid's variation cycle, the outer shell of the atmosphere contracts. The temperature of the star's surface increases, and the star's output reaches its maximum level.

In phase 3 the outer atmosphere expands once more, the surface temperature decreases, and the light output reduces to its minimum.

HENRIETTA SWAN LEAVITT

In 1912, American astronomer Henrietta Swan Leavitt (1868–1921) discovered Cepheid variables in the Small Magellanic Cloud. It was possible to correlate the period of some of these with their apparent magnitude. Since they were all in the Magellanic Cloud – and thus at roughly the same distance from Earth – their period could be fixed to their luminosity. So from a Cepheid variable's period, its absolute magnitude can be calculated. Its apparent magnitude can be measured from Earth, making it possible to figure out how far away it is. This makes Cepheids very useful measuring rods. By timing the period of the star's variation cycle, the distance of the star from Earth can be calculated using a simple formula.

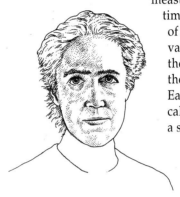

inside the star, like water dammed up behind an obstruction, before slowly blowing the soot away so the output can increase again.

Some variable stars erupt or flare up: their outputs suddenly increase by three magnitudes. This can happen several times in 24 hours. These stars are associated with nebulae, and they seem to be young stars in the process of forming. Some of them flare up irregularly, others regularly. They seem to oscillate like a spring, regularly expanding and compressing their outer layers.

Variables can also be stars that are part of a binary system. In fact, the most common type of variable is the eclipsing binary, in which closely orbiting stars pass in front of each other, thus altering their combined light output, but in a regular way.

Building solar systems

Stars and their attendant solar systems form from the most insubstantial materials – tenuous gas and dust.

The world on which we stand and the Sun that shines on it have existed for a long time – about 4.5 billion years. However, there have been stars shining for perhaps 13 billion years, and many of them exploded before our sun was born. So how do stars come into existence, and what raw materials are needed to make a sun and planets like ours?

The first generation of stars, which formed about 2 billion years after the Big Bang, were made of just two elements: hydrogen and helium. By that time, the universe had cooled down somewhat, and the clouds of hydrogen and helium gas left behind by the Bang had collected in and around galaxies. In some parts of the clouds, matter was concentrated, locally raising the gravitational force. These concentrations thus slowly attracted further matter, and the clouds began to collapse. This made the denser areas even more dense, pulling in yet more matter and speeding up the collapse.

Eventually, matter was forced by gravity into a small volume, while pressure caused the hydrogen gas to condense into liquid. As the ball of liquid hydrogen grew, the center got more compressed, the internal pressure increased, and the liquid hydrogen became a metal-like solid. In any ball of hydrogen bigger than about one-sixteenth the size of the Sun,

When an ice skater spins (above), her rotation rate is constant for any given body posture. If she draws in her arms, say, the spin speeds up because the momentum in her arms is passed on to her body.

Something similar happens in a spinning gas cloud. When it is large, the spin rate is slow, but the rate speeds up as the cloud contracts when it collapses in the process of forming stars.

As a cloud of interstellar gas and dust contracts under the force of gravity (**1**), the spinning mass forms a disk, with a bulge in the center where a warm protostar forms (**2**). The center collapses under its own gravity and continues to heat up while the gas and dust in the surrounding disk continue to fall in. The protostar radiates much heat and ejects matter outward from its poles – where it is not obstructed by the disk – getting rid of much gas and

dust in the process (**3**). Eventually, fusion starts in the core (**4**), and the star starts its active nuclear life. Only stars that are 6 percent or more of the mass of the Sun can attain the temperature and pressure in the core that are required to initiate fusion. The disk, meanwhile, either disperses or forms planets.

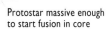
Protostar massive enough
to start fusion in core

Outflow sweeps away much gas and dust
surrounding protostar

Protostar forms
with warm core

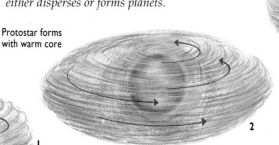

Cloud of gas shrinks
and starts to spin

6

Outer planets retain light gases

Light gases driven off inner planets

Matter swirling around a new star
forms small pellets which collide to
make larger bodies called planetoids
(5). These coalesce, forming large
planets with stretches of mostly
empty space between them. In the
inner system, light gases are blown
off by the star's radiation (6),
leaving rocky planets behind.

5

Gas and dust create planetoids around star

The star Beta Pictoris (below)
has a broad disk of dust around it
that stretches for many tens of billions
of miles, as revealed in this infrared image.
This disk, which contains some carbon-based
chemicals, could be a solar system in the making.
It has been estimated that about one-third of the newly
forming stars with the same mass as the Sun or less have
the right conditions for the formation of planets.

MAKING PLANETS

No one knows how many stars have planets orbiting them. First-generation stars, which form from hydrogen and helium only, might have planets orbiting them, but these could only be gas planets like Jupiter, but with no rocky core. For Earthlike planets to form, the star must be a second- or third-generation star made of a hydrogen and helium cloud laced with heavier elements.

Since our sun is fairly typical, it seems unlikely that other stars do not have similar solar systems. Today's telescopes are not powerful enough to see directly whether this is the case. However, our near neighbor, Barnard's star, wobbles as it moves across the sky. Calculations show that the wobble could be caused by the gravitational effects of two Jupiter-sized planets.

the force of gravity holding it together makes the temperature and the pressure at the center high enough to set off a fusion reaction. Hydrogen atoms are forced together to make helium, giving off energy, and a star is born.

When some of these first stars had finished processing the hydrogen, they began fusing the helium into carbon; and in really large stars, further, heavier elements were manufactured. The lives of some of these first-generation stars ended in huge explosions, called supernovae, which blasted heavy elements into space. These enriched the interstellar gas which formed the next stars.

When the gas and dust cloud that forms a second- or even third-generation star – like our sun – starts to collapse, there are tiny amounts of heavier elements in it. These heavier elements, including iron, silicon, and oxygen, have formed the Earth, while carbon makes life possible. Since these elements formed in distant stars, the Earth and all life on it are, in fact, recycled stellar debris.

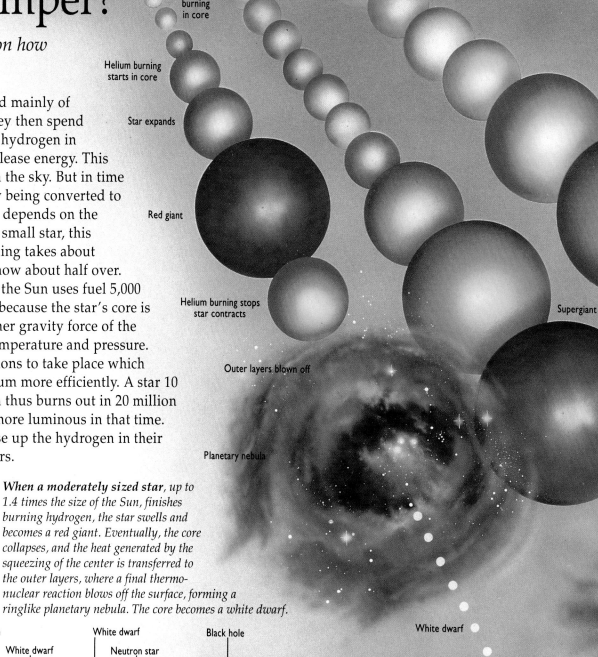

1 solar mass   10 solar masses   30 to 50 solar masses

Steady hydrogen burning in core

Helium burning starts in core

Star expands

Red giant

Helium burning stops star contracts

Outer layers blown off

Planetary nebula

Supergiant

White dwarf

Bang or whimper?

A star's eventual fate depends on how massive it was to start with.

Stars start their existence composed mainly of the simple element hydrogen. They then spend most of their active lives consuming hydrogen in their cores in fusion reactions that release energy. This is what makes them shine steadily in the sky. But in time the hydrogen fuel will be used up by being converted to helium. How long that process takes depends on the star's size. In our sun, a relatively small star, this steady phase of hydrogen burning takes about 10 billion years. That time is now about half over.

A star 10 times the mass of the Sun uses fuel 5,000 times more quickly. This is because the star's core is squeezed more by the higher gravity force of the extra matter, raising its temperature and pressure. This permits fusion reactions to take place which convert hydrogen to helium more efficiently. A star 10 times the mass of the Sun thus burns out in 20 million years and is 5,000 times more luminous in that time. The most massive stars use up the hydrogen in their cores in a mere million years.

0.1 solar mass

Brown dwarf

Cooling down

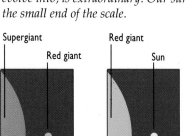

Tiny stars (left), with less than 0.08 times our Sun's mass, glow dimly as brown dwarfs, then fade slowly to become black dwarfs.

The difference between the sizes of the biggest and smallest stars, and what they evolve into, is extraordinary. Our sun is at the small end of the scale.

When a moderately sized star, up to 1.4 times the size of the Sun, finishes burning hydrogen, the star swells and becomes a red giant. Eventually, the core collapses, and the heat generated by the squeezing of the center is transferred to the outer layers, where a final thermonuclear reaction blows off the surface, forming a ringlike planetary nebula. The core becomes a white dwarf.

Supergiant
Red giant

Red giant
Sun

Sun
White dwarf

White dwarf
Neutron star

Black hole
Neutron star

Stars about 10 times the mass of the Sun are blue initially, but then enlarge. They become red supergiants when the fusion reactions change as the hydrogen runs out in their cores. Eventually, the nuclear reactions cease, and the cores collapse, causing the stars to blow apart, resulting in supernovae. Some explode so violently that they spew out much of their mass into space. They leave behind tiny, extremely dense objects called neutron stars.

The largest stars, with a mass about 30 times that of the Sun, really do live fast and die young. They leave, if not a beautiful, then at least an interesting corpse. These massive stars end in extremely powerful supernovae explosions. Their cores are so massive that they continue to collapse past the neutron star stage to become black holes.

CHEMISTRY IN THE CORE

It is only the thermal, or heat, energy generated by the nuclear furnace in the center of a star that prevents it from collapsing under its own gravity. Once the hydrogen in a star's core has turned into helium, the nuclear reaction in which hydrogen is fused ceases. But if the star is massive enough, the core starts shrinking, which generates heat.

When the heat reaches the outer part of the star, it sets off a new fusion reaction in the hydrogen there. The heat being generated makes the outer regions of the star swell up, and the star gets brighter and larger. The new fusion reaction makes more helium, which falls to the core. Although the core becomes more massive, it continues to shrink under its own weight. Eventually, this contraction puts the temperature up to more than 100 million K. At this point, helium starts to fuse to produce the stable element carbon. Eventually, carbon chokes the core.

In even larger stars, further collapse increases the temperature, and the carbon fuses to form magnesium or neon. Carbon also fuses with helium to create oxygen which, in addition, can be fused to give silicon or sulfur. Yet more, even heavier elements are formed as these larger nuclei are bombarded with free neutrons released by the various fusion reactions.

Thermonuclear reactions continue making various new elements until the core is choked with iron. Then the star is ready to become a supernova, because when iron undergoes fusion, it takes in rather than gives out energy. With no energy being made in the core, the star collapses and the outer layers are hurled out in a titanic explosion.

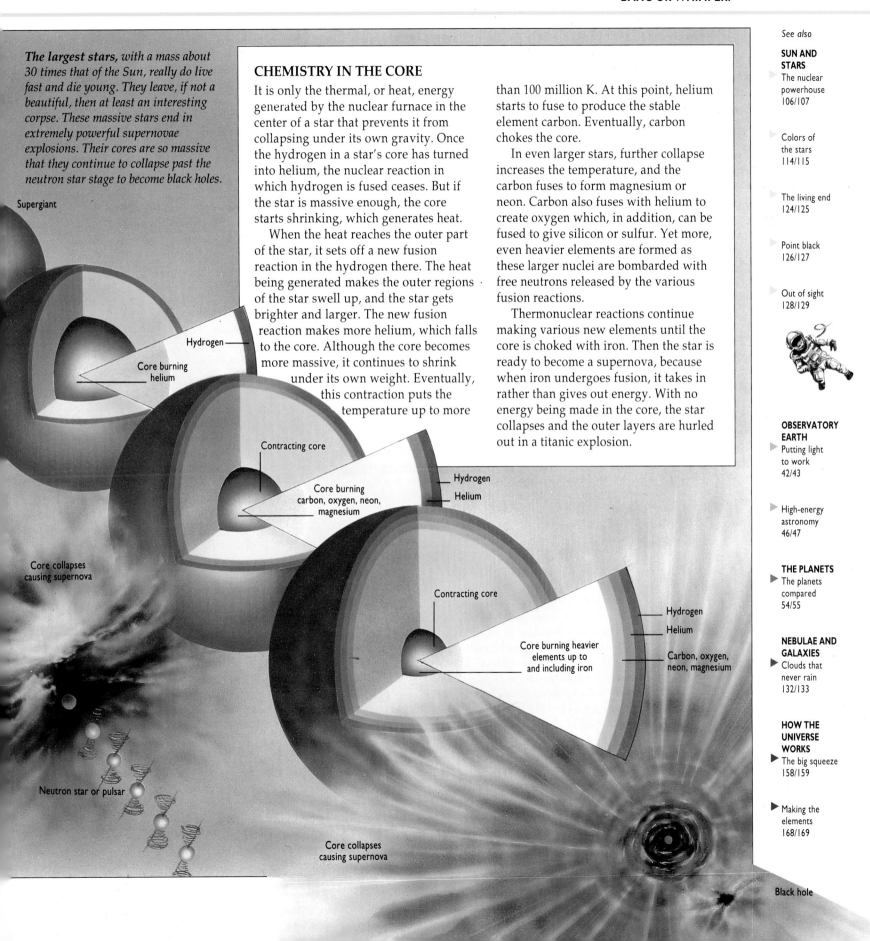

Supergiant

Hydrogen

Core burning helium

Core collapses causing supernova

Neutron star or pulsar

Contracting core

Core burning carbon, oxygen, neon, magnesium

Hydrogen

Helium

Contracting core

Core burning heavier elements up to and including iron

Hydrogen

Helium

Carbon, oxygen, neon, magnesium

Core collapses causing supernova

Black hole

The living end

Death is a crushing blow for most stars, but some leave behind a pulse.

After the fireworks that can mark the death of a star, one of four fates awaits what remains. A small brown dwarf just gets browner and fades away. A star about the size of the Sun blows off its outer layers, leaving behind a small white dwarf in which matter is squeezed and compressed. A star somewhat larger than the Sun goes out in style with a supernova explosion. An incredibly compressed neutron star – in which matter is scrunched down to occupy the smallest possible space – remains as its celestial gravestone. The biggest stars make the biggest supernovae and leave mysterious, light-hungry points of nothingness – black holes.

White dwarfs are the remnants of stars up to 1.4 times the mass of the Sun. They are small, about the same size as

Beam of radio emission

JOCELYN BELL BURNELL

In 1967, at the Mullard Radio Astronomy Laboratory in Cambridge, England, a post-graduate student, then named Jocelyn Bell (1943–), found a strange signal while making a radio sweep of the sky. It pulsed regularly every 1⅓ seconds.

At first, Bell and her colleagues thought they had stumbled across radio signals from aliens. Later that year, British cosmologist Thomas Gold suggested that a "pulsar" could be a spinning neutron star – the collapsed core of a supernova – so it was logical to look for them in supernova nebulae. A radio pulsar that coincided with the position of a star flashing in visible light was found in the center of the Crab nebula, the remnant of a supernova seen on Earth in A.D. 1054, and the theory was proved. Hundreds of pulsars are now known.

Axis of rotation

Magnetic force lines

Path of radio beam

Photosphere

External crust

Solid core

Internal crust

Neutron fluid region

Magnetic force lines

A neutron star has a core of neutrons around which is a fluid layer made up of neutrons, protons, and electrons. Then there is a thin, two-part crust. The original rotation energy of a sun is preserved in a neutron star – the compressed core it leaves behind when it becomes a supernova – so neutron stars can spin fast, at between 1,000 times a second and once every few seconds. A star's original magnetic field is also condensed in a neutron star, making it extremely intense. Charged particles – protons and electrons – given off from the surface of the neutron star are caught up in the magnetic field and spun around to emit radio waves in narrow beams at the magnetic poles of the star. If, as is often the case, the axis of rotation does not coincide with the axis of the magnetic field, the radio beams will sweep across the sky, delivering short pulses of radio waves to any receiver in their path.

Beam of radio emission

*The remnants of a supernova explosion are visible as the glowing gas and dust that can be seen spread out across space in images from long-exposure photographs taken through telescopes. The material (**left**) is the remains of the Vela supernova, where a massive star exploded at the end of its fusion-fueled life about 10,000 years ago.*

While the outer layers of the star were sent hurtling into space, the core collapsed in on itself, forming a small, incredibly dense neutron star. The Vela and Crab nebula remnants are the only two supernova sites definitely to have associated central pulsars. The pulsar in the Vela remnant sends out a radio pulse every 90 milliseconds.

DEGENERATE MATTER

Everyday matter is largely empty space: in an atom, a tiny nucleus made of protons and neutrons is surrounded by even tinier orbiting electrons which are, compared with the size of the nucleus, a long way away.

When matter is crushed by the enormous pressure in a white dwarf, it collapses only so far before it becomes "electron degenerate." What this means is that the free electrons in the gas of the star are crammed together so much that they occupy positions close to atomic nuclei, although atomic nuclei can still move about freely. The packed electrons resist further crowding with such enormous force that the star stops collapsing.

If the pressure is higher, as in the collapsing core of a supernova, electrons are forced to combine with protons to make neutrons. The matter then reaches a lower stage of degeneracy in which it is made entirely of neutrons, as in a neutron star. Neutrons resist further crushing with fantastic force. If the force is extremely strong, however, matter degenerates completely, collapsing into a singularity – a black hole.

A planetary nebula shows where a Sun-sized star has died and blown off its outer layer. At the center of the Helix nebula, NGC 7293 (**below**), is a white dwarf of degenerate matter.

the Earth, but 10,000 times heavier. All the matter in them has been so crushed that one cubic centimeter would weigh a ton. The surface temperature of a white dwarf is high; at 17,000K it is around three times the temperature of the surface of the Sun. Inside, however, a white dwarf is relatively cold – less than 1 million K, compared with the Sun's core temperature of 10 million K. This is too cool for any fusion reactions to take place, so the white dwarf radiates off its remaining energy and slowly dims over the course of billions of years. Eventually, all that is left is a black cinder of degenerate matter. The outer atmosphere blown off from the star forms an expanding ring, which glows as a result of energy it receives from the central white dwarf.

After its supernova explosion, a star more than 1.4 times the mass of the Sun leaves behind a large core of solid iron. With no fusion reactions taking place, there is nothing to keep the core from collapsing. Since the core remnant is larger than that of a white dwarf, the force of gravity is even stronger. It presses the remaining matter into an even smaller space, where it degenerates further until only crammed-together neutrons (subatomic particles) remain. The result, a neutron star, is very small, only about 20 miles (32 km) across, and the matter in it is extremely dense – one cubic centimeter would weigh 1 billion tons. The matter blown off in the supernova explosion forms an expanding cloud that gradually disperses, becoming part of the gas and dust of the interstellar medium.

Point black

Some dying stars collapse to become black holes and vanish from sight.

The cores of the most massive stars – those more than 30 times the mass of the Sun – collapse dramatically when the fusion reactions that power them cease. In a star this size, gravity pulls matter inexorably inward so that the core quickly passes through the white dwarf and neutron star stages. Then something extraordinary happens: it keeps on collapsing until it disappears into a point infinitely smaller than the period at the end of this sentence. It becomes a massively heavy, infinitely dense dimensionless object called a singularity.

The gravitational field around a singularity is so strong that nothing, not even light, can escape. Indeed, light is not only unable to shine out from a singularity, it cannot be reflected from one either. Even light that happens to be going past is sucked in, and anything that falls in is lost forever. Because of these properties, the space around a singularity where light cannot escape is known as a black hole. While a singularity has no dimensions, its black hole has a size.

The more massive the singularity, the wider the region around it

When the core of a star starts to collapse, light can escape from its gravitational field normally (**1**). As the collapse proceeds, the field gets stronger and some light is bent back (**2**). There comes a point during the collapse when only light emitted vertically from a star can escape (**3**). When a star collapses below a certain size (the Schwarzschild radius), light from it cannot escape and it becomes a black hole (**4**). Matter in the star is crushed down below the smallest size that it can theoretically occupy, forming a singularity (**5**).

The Schwarzschild radius is also the star's event horizon, the region in which events inside the star cannot be detected, and it defines the size of its black hole. The black hole's width depends on the singularity's mass. In theory, it is possible for small black holes to exist. Calculations show that a singularity with the mass of the Earth would have a black hole 1 inch (2 cm) across. A stellar black hole would be only a few miles across. A black hole the mass of several hundred million stars would be the width of our solar system.

where light cannot escape and thus the wider its black hole.

Other regions where black holes are thought to exist are at the centers of some galaxies. Movements of gas and dust clouds in the center of our own galaxy, for instance, suggest that there is a singularity there with the mass of two to three million stars. It resides in the center of a black hole no bigger than the diameter of our sun. There is also thought to be a huge object, possibly a singularity, the mass of 30 to 70 million Suns in the middle of M31, the galaxy in Andromeda.

Light escapes from star's low intensity gravity field

Star begins to collapse

Schwarzschild radius

Light rays closest to surface bent back by gravitation

Star collapse accelerates

Just before reaching Schwarzschild radius, only vertical light rays can escape

No light can escape once star reaches Schwarzschild radius

Singularity

Event horizon

Entire mass of star collapses into black hole

Around a black hole, the gravity field can be shown as a grid (**left**) stretched by the mass of the singularity. For a star-sized object, the grid is slightly indented. For a black hole, it stretches down out of sight.

Light itself can bend under the influence of gravity. The gravity field on Earth is so slight that, to all intents and purposes, light travels in a perfectly straight line. Even the Sun, the most massive nearby object, bends light passing close by it from distant stars only by a tiny fraction of a degree. Only when much mass is concentrated in small volumes does light bending becomes more pronounced. The infinite density of a singularity in a black hole provides the ultimate gravity field. If light is bent by gravity, or any other means, objects out of the line of sight become visible. A good way of demonstrating this effect is with a photograph taken with a camera using a fish-eye lens, such as this image taken in the Rockefeller Rink, New York City. Light from objects all around seems to crowd in on the field of view. Being in a black hole would be like being at the bottom of a light well. Instead of the the light coming in from only 180°, as in a fish-eye lens image, it would come into the black hole from all around. The view of the outside would be distorted but all encompassing. If you could survive in a black hole, you would be able to look ahead and see behind you at the same time.

Detecting a black hole might seem impossible: if no light or other radiation emerges from it directly, what is there to detect? There are ways though. Black holes suck matter in from space around them, and as this material is sucked toward the black hole, it spirals in emitting X-rays. These are in the form of brief, hard-to-detect flashes of radiation given out just before the matter disappears into the hole.

If a black hole is part of a binary star system, however, it should slowly suck away the atmosphere of its companion star, producing strong, continuous X-ray emissions that should easily be detected. Several such sources, including Cygnus X-1 (**right**), have been found, and they are prime black hole candidates.

Out of sight

At the edge of darkness that defines a black hole, intense gravity plays some bizarre games with the laws of physics.

A black hole's event horizon – the boundary beyond which even light cannot escape – is the point of no return. Science predicts some weird happenings near the horizon of a black hole and beyond.

Long before the existence of black holes was suggested, Albert Einstein pointed out that space and time were intimately connected. Inside a black hole, and in the area around it, space is stretched out and time is slowed down by the effects of the enormous gravitational field.

If you were outside a black hole and saw an astronaut fall in, you would, in fact, never see him reach the event horizon. As he neared the edge of the black hole, time would slow, and he would seem to be traveling more and more slowly. This time slowing would have an effect on the wavelength of the light coming from him, and the astronaut would appear redder and redder until he faded away.

For the unfortunate astronaut, time in the universe outside would speed up. But at some point he would be torn apart by the gravitational field of the black hole. If he fell in head or feet first, he would be stretched out – "spaghettified" – because the pull of gravity on one end of his body would be so much stronger than the pull of gravity on the other end. Eventually his body would no longer be able to stand the stretching and he would be ripped apart.

Exactly when this happened would depend on the size of the black hole. With a small black hole, the astronaut would probably be spaghettified before he reached the event horizon. But with a big black hole, he might survive inside for a few hours or even days.

Racing toward the event horizon of a black hole, a person would seem, to an observer watching from a safe distance, to take the same amount of time to cover less and less distance the closer he got to the edge. He would thus seem to move more and more slowly until he became infinitely slow, just before he reached the event horizon. The effect is like a TV replay of a sporting event in which the action is gradually slowed down until an athlete is stopped at the moment of reaching the finishing line, never to cross it.

The person approaching the black hole would, conversely, perceive time to be passing normally, but the universe around him would appear to be speeding up.

Past a certain point, a person approaching a black hole would be "spaghettified" (**far right**) by the difference between the gravitational force at each end of his body, and then torn to shreds.

At the event horizon of a black hole that contained 10 times the mass of our sun in its central singularity (the infinitely dense remains of a large collapsed star), a person approaching feet first would feel the same stretching force as if he were on Earth, but with the population of Paris (about 8.5 million people) hanging from his ankles (**right**). The person would have been torn apart, however, about 250 miles (400 km) from the event horizon.

To an outside observer, a person approaching black hole appears to slow down

Body approaching black hole is stretched by differences in gravity

Event horizon

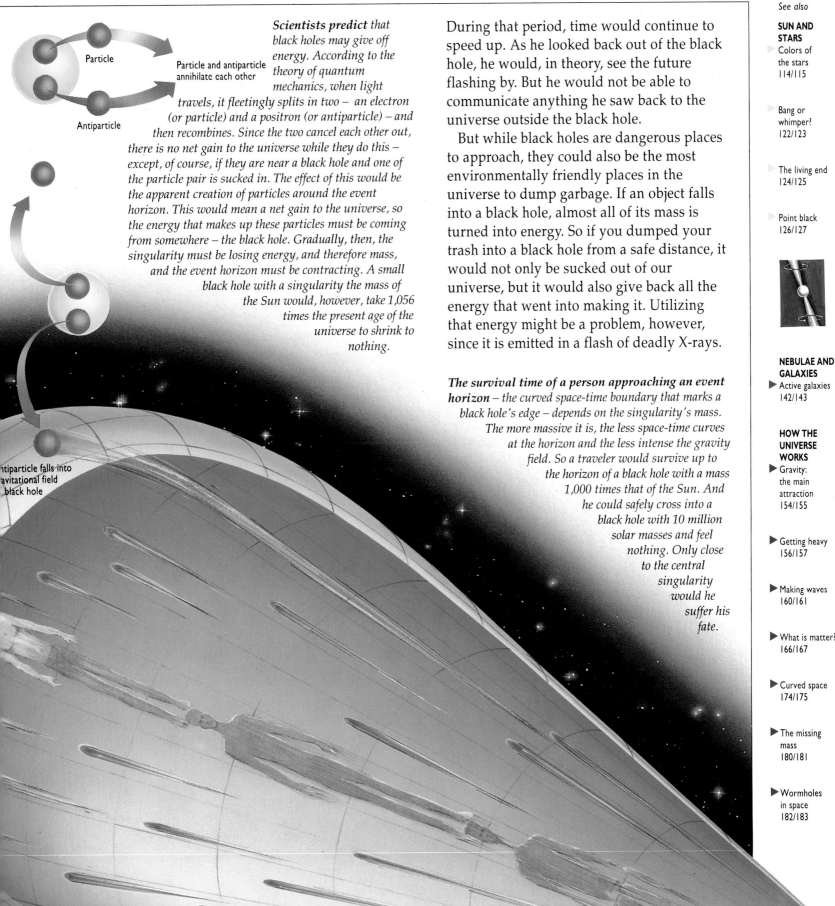

Particle

Antiparticle

Particle and antiparticle annihilate each other

ntiparticle falls into avitational field black hole

Scientists predict that black holes may give off energy. According to the theory of quantum mechanics, when light travels, it fleetingly splits in two – an electron (or particle) and a positron (or antiparticle) – and then recombines. Since the two cancel each other out, there is no net gain to the universe while they do this – except, of course, if they are near a black hole and one of the particle pair is sucked in. The effect of this would be the apparent creation of particles around the event horizon. This would mean a net gain to the universe, so the energy that makes up these particles must be coming from somewhere – the black hole. Gradually, then, the singularity must be losing energy, and therefore mass, and the event horizon must be contracting. A small black hole with a singularity the mass of the Sun would, however, take 1,056 times the present age of the universe to shrink to nothing.

During that period, time would continue to speed up. As he looked back out of the black hole, he would, in theory, see the future flashing by. But he would not be able to communicate anything he saw back to the universe outside the black hole.

But while black holes are dangerous places to approach, they could also be the most environmentally friendly places in the universe to dump garbage. If an object falls into a black hole, almost all of its mass is turned into energy. So if you dumped your trash into a black hole from a safe distance, it would not only be sucked out of our universe, but it would also give back all the energy that went into making it. Utilizing that energy might be a problem, however, since it is emitted in a flash of deadly X-rays.

The survival time of a person approaching an event horizon – the curved space-time boundary that marks a black hole's edge – depends on the singularity's mass. The more massive it is, the less space-time curves at the horizon and the less intense the gravity field. So a traveler would survive up to the horizon of a black hole with a mass 1,000 times that of the Sun. And he could safely cross into a black hole with 10 million solar masses and feel nothing. Only close to the central singularity would he suffer his fate.

Nebulae and Galaxies

*S*plashed across the heavens is a faint band of pearly luminescence, visible with the naked eye only when the sky is dark and clear. This is the Milky Way, and legend has it that it is the path to the palace of the Greek god Zeus. It is, in

fact, made of the blended light of a vast collection of stars so distant that they are individually indistinguishable. These stars are those of the Galaxy, the spiral-shaped eddy of stars around which our sun slowly orbits and of which we are a part. Our galaxy is also home to star clusters and clouds of gas and dust, called nebulae. Beyond it are other galaxies; the closest can be seen as dim pools of light, but most are observable only with powerful telescopes. These other galaxies stretch for billions of light-years beyond our galaxy in clusters and mighty superclusters.

Left (clockwise from top): a luminous gas cloud; a galaxy evolving; the Pleiades, a star cluster; matter swept together to form stars; our galaxy. This page (top): a spiral galaxy made of stars, gas and dust; (left) colliding galaxies spewing star streamers into space.

Clouds that never rain

Space is far from just an empty vacuum. Scattered through it are vast clouds of gas and dust.

In addition to the pinpoints of the stars, there are about two dozen dim, fuzzy patches of light, like little glowing clouds, visible to the naked eye in a clear sky. Long-exposure photographs taken through large telescopes reveal that there are thousands upon thousands of these blurs of light in the heavens. We now know that these patches of brightness include faint clusters of stars and distant galaxies, as well as wisps of glowing gases and banks of dust glistening in starlight. But until it was shown in the 1920s that many are really galaxies, all of them were called nebulae, Latin for "clouds." Only shining gas and dust deserve the name nebulae, however, as they are truly clouds.

Also present in the heavens are many invisible clouds that can be detected only by the radiation they emit or by their effect on light passing through them. These, along with the visible clouds, are known as the interstellar medium and tend to be concentrated in the spiral arms of galaxies. Constantly changing as clouds drift apart or clump together, the interstellar medium is 99 percent hydrogen and helium gas with a tiny amount of other gases and even less icy cosmic dust.

The Horsehead nebula (right) in Orion appears as a distinctive silhouette against the brighter emission nebula behind. It stands out from the uniform background because it is an area with particularly dense dust concentrations. These absorb the light emitted from the regions beyond, giving the cloud its dark appearance.

A wider view of the Horsehead nebula reveals the amount of dust found in the gas clouds in this region of the sky. The photograph shows that the "horse's head" is merely a part of a much larger cloud beneath. A comparison of the number of stars visible in the upper and lower halves of the picture shows that fewer stars appear in the lower half. This is because light from behind is blocked out by the dust in this large dark nebula. Only stars that are in front of the cloud can be seen.

The horse's head protrudes into a bright red strip of ionized hydrogen gas. The gas has been made to glow its typical pink color by light from Sigma Orionis, the bright star at the top of the picture.

The bright patch just below and to the left of the horse's head, NGC 2023, is a reflection nebula, reflecting the light from stars within the dusty cloud. The bright star on the left of center is Zeta Orionis, or Alnitak. This region actually forms part of the enormous cloud of gas which also includes the much brighter Orion nebula.

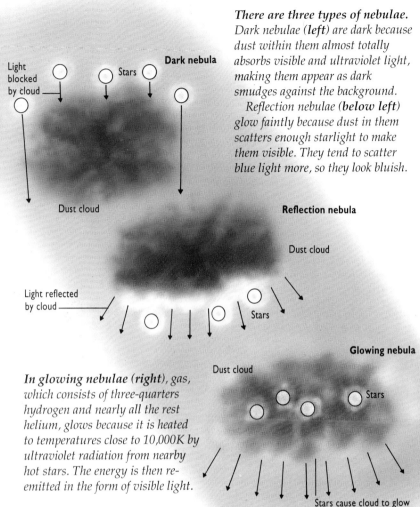

*There are three types of nebulae. Dark nebulae (**left**) are dark because dust within them almost totally absorbs visible and ultraviolet light, making them appear as dark smudges against the background.*
*Reflection nebulae (**below left**) glow faintly because dust in them scatters enough starlight to make them visible. They tend to scatter blue light more, so they look bluish.*

Light blocked by cloud
Stars
Dark nebula
Dust cloud

Reflection nebula
Dust cloud
Light reflected by cloud
Stars

Glowing nebula
Dust cloud
Stars
Stars cause cloud to glow

*In glowing nebulae (**right**), gas, which consists of three-quarters hydrogen and nearly all the rest helium, glows because it is heated to temperatures close to 10,000K by ultraviolet radiation from nearby hot stars. The energy is then re-emitted in the form of visible light.*

CHARLES MESSIER – CATALOGING THE CLOUDS

Nicknamed the comet ferret by Louis XV, Charles Messier (1730–1817) was a French astronomer, born in Badonvillier. He was a keen comet-hunter and, to eliminate any objects that might mistakenly be identified as comets, he made a list of all the vague blurs in the sky that he could see using the instruments available to him. In 1771 he published a list of 45 such objects, which he had extended to 103 by 1784. In 1789 his compatriot, Pierre Méchain, then added a further 6 to bring the total to 109. These objects were identified in Messier's catalog with a number, now known as the M, or Messier, number. For example, M31 is a galaxy in the Andromeda constellation, while the Orion nebula is M42. However, a couple of original objects identified by Messier have since gone "missing," such as M47 and M91. It may be that they really were comets. Since Messier's death, many astronomers have extended the catalog of these vague objects in the sky. The Danish astronomer J.L.E. Dryer compiled his New General Catalog (NGC) in 1888. The list included all of Messier's objects, and as in Messier's list, they were given a number, their NGC number. Hence M1, the Crab nebula, is also known as NGC 1952, and M31, the galaxy in Andromeda, is also NGC 224.

The star factories

Vast clouds of gas scattered through the Galaxy are the raw materials from which stars are made.

The most massive objects in galaxies are not stars, but huge clouds of interstellar gas and dust called giant molecular clouds, often up to 200 light-years across, and containing a million times as much matter as the Sun. They are made almost entirely of two gases (hydrogen and helium) and some dust, like other interstellar clouds. These clouds are so cold – just 10K above absolute zero (which is the coldest possible temperature) – that within them some atoms bind together to form molecules. The hydrogen atoms thus link in pairs to form H_2 molecules. Helium atoms are stable and stay single.

Over 60 different molecules have been identified within these clouds. The two most important are hydrogen and carbon monoxide, which are both found in greatest abundance in the most dense clouds. The molecules detected range from simple ones, such as water and ammonia, to more complex molecules such as formaldehyde and ethyl alcohol.

Some clouds also seem to contain molecules such as acetaldehyde, which is one of the components of amino acids that are essential for life.

Matter is not evenly distributed within the clouds, but is clumped together in places. These clumps can grow more dense and eventually become the birthplace of stars. The size, temperature, and molecular content of the cloud determines the size and life history of the stars that are destined to be born within it. Molecular clouds are found in the younger areas of a galaxy, that is, regions where star birth is continuing today. In a spiral galaxy, like our Milky Way, this means the spiral arms. On close inspection, the spiral region of the galaxy in which the Sun is found reveals many such clouds, including the Orion nebula.

Shock waves from the explosion of giant stars are thought to sweep through giant molecular clouds like a cosmic broom (left), pushing gas and dust closer together. The increased mutual gravitational attraction between the particles then begins to pull all the material together until eventually a star is formed.

134

STAR BIRTH IN THE CLOUD CRADLES

Because particles are unevenly distributed within a molecular cloud, areas develop in which matter is concentrated. More particles gradually join these clumps, creating fairly dense regions. Clumping may be aided by disturbances such as the shock waves of expelled matter from a nearby supernova. As more and more molecules condense, the area's gravity becomes so strong that it begins to collapse.

These very dense regions, or Bok globules, have a mass of more than 200 times that of the Sun, and measure about 3.5 light-years across. As the globule continues to condense into a protostar, the temperature rises. When it reaches 10 million K, thermonuclear reactions begin and the star ignites, sending shock waves rippling through the cloud.

From observations of molecular clouds like the one in Orion, it is known that star formation tends to sweep slowly through a giant molecular cloud from one side to another. In the Orion nebula, star formation began about 12 million years ago at the eastern edge and has proceeded gradually westward. Stars in Orion's belt are about 8 million years old and get progressively younger toward the Trapezium cluster. Here, stars are forming at this moment.

The march of star formation through a molecular cloud can be initiated by a massive star at one edge. When a big star burns out, it explodes as a supernova, and the energy of the explosion ripples through the surrounding gas (**1**). These ripples, in the form of fast-moving matter hurled out by the explosion, cause disturbances within the cloud so that molecules become

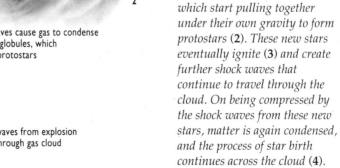

Younger stars form on subsequent shock waves

4

New stars ignite causing further shock waves

3

Shock waves cause gas to condense into Bok globules, which become protostars

2

concentrated in certain areas. As the material is squeezed together, it becomes compressed enough to form dense cores of material which start pulling together under their own gravity to form protostars (**2**). These new stars eventually ignite (**3**) and create further shock waves that continue to travel through the cloud. On being compressed by the shock waves from these new stars, matter is again condensed, and the process of star birth continues across the cloud (**4**).

The Cone nebula in Monoceros (*above*) is the scene of star birth today. There are many young, bright T Tauri stars nearby which vary in brightness because they are still surrounded by swirling remnants of the gas and dust from which they formed. The nebula itself is set glowing by this cluster of hot stars. It owes its pinkish color to the high percentage of hydrogen gas within the cloud.

Supernova explosion

Shock waves from explosion ripple through gas cloud

I

Star clusters

Rarely found completely alone in galaxies, stars are often grouped together in bright clusters.

At certain places in the night sky, there are bright patches of light millions of times brighter than the Sun – although they are so far away that they appear quite dim when seen from Earth. These bright patches are clumps of stars and are known as globular clusters because they hang in space like globules, forming nearly symmetrical balls of stars. Near the center, these clusters appear so densely packed that individual stars cannot be distinguished. But it is an illusion – even here the stars are many light-months apart.

Globular clusters are enormous structures, typically 100 light-years across. Although it is only possible to count up to 30,000 or so individual stars in the nearest cluster, they all probably contain hundreds of thousands. However, most of these are too dim to see. The most luminous stars in globular clusters are reddish in color, and astronomers have deduced that the brightest of these must be red giants.

There are about 140 globular clusters associated with our galaxy, most of them gathered in a halo around the flat disk of the Milky Way. They are among the oldest structures in the universe, having formed more than 10 billion years ago, and contain only ancient stars.

M5, also called NGC 5904 (left), is a large globular cluster. The stars within this cluster, as with all globular clusters, are very old. The biggest stars have burned up their supply of hydrogen to become red giants, or aged even further to become white dwarfs or neutron stars. Only the smallest stars, less massive than the Sun, are still burning normally.

The Pleiades (above) is the best known of all open clusters. With the naked eye, it appears to be a small group of six or so stars in the shape of a plow in the constellation of Taurus. Through a telescope, it is clear that it actually contains hundreds of stars.

Photographs show that the stars are surrounded by wisps of gas that reflect the light of the stars.

The stars of open cluster NGC 6520 (below) are a remarkably intense blue. Close by it is a dark nebula made of opaque gas and dust that blocks out light from stars beyond. The intense blue color indicates that the stars are very young. This is because stars can only be this color when they are massive, and massive stars burn out quickly, in the course of a few million years. By astronomical standards, this cluster has emerged from its gas cloud relatively recently.

After a cluster has formed, the stars gradually move apart. Even the Sun may have possibly begun its life in a cluster, but over time the individual stars have dispersed.

Globular clusters do not sit within the flat disk of the Galaxy, but lie on orbits at an angle to it. They ring the Galaxy, slowly orbiting around it far from the center, taking about 100 million years to complete a circuit.

Besides these old clusters, there are other, much younger groups of stars, called galactic clusters, visible within the disk of the Milky Way. Some galactic clusters are just a few million years old. There are thousands of these clusters – often known as open clusters because they form a loose group – but they are tiny compared to globular clusters. Most contain no more than 100 stars, and each star is discernible through a powerful telescope.

There are also even younger groups of stars, called associations, scattered along the spiral arms of the Galaxy. These associations are linked with the vast interstellar clouds where stars are born, and they contain some of the youngest of all stars, including highly luminous, short-lived blue stars.

All clusters have NGC (New General Catalogue) numbers. Some also have M (Messier) numbers, after the 18th-century French astronomer Charles Messier who cataloged many clusters and other objects.

NGC 3293 (right) is an open cluster in the constellation of Orion. The cluster is between 5 and 10 million years old. All the stars within it are young, and only the most massive have evolved to the stage where they are red in color. There is just one bright orange star in this cluster, and it is clearly extremely luminous, having swollen up to enter its supergiant phase prior to ending its life as a supernova.

The youngest stars are on one side of this cluster, while the oldest are on the opposite side. This supports the theory that star formation sweeps across a gas cloud from which a cluster forms, rather than happening at the same time throughout a cloud. A wave of formation comes about because when stars ignite they send out shock waves. These compress gas which then collapses to form stars. New stars on one side of the gas set up a chain reaction which spreads across the cloud, creating stars as it goes.

Islands in space

In the void of space, stars are bunched together into galactic islands, millions of light-years apart.

Most of the smudges of light visible through telescopes are not nebulae – clouds of gas and dust – but separate galaxies beyond our own Milky Way. Only long exposure photographs reveal their often intricate structures. The biggest giant elliptical galaxies contain thousands of billions of stars and may be several hundred thousand light-years across. Even the tiniest "dwarf" galaxies contain millions of stars. The nearest large galaxy, the Andromeda galaxy, is over 2.2 million light-years away. In 1936 Edwin Hubble published a system for galactic labeling. He divided galaxies into four different classes according to their shape. These are spiral (S), barred spiral (SB), elliptical (E), and irregular galaxies.

Spiral galaxies are labeled at the top: Sa, Sb, Sc

Elliptical galaxies E0, E3, E7, S0

Spiral galaxies, like M100 (**right**) and our own Milky Way, are among the most extraordinary structures in the universe. Like giant spinning firework wheels, they have a central core and vast spiraling arms containing millions upon millions of stars of all ages. These galaxies are classified according to the size of their central region and the openness of the arms. Sa types have large nuclei and tight spiral arms, Sb types have smaller nuclei and less tightly wound spiral arms, and Sc types have the smallest nuclei and the most unwound spiral arms.

Neither elliptical nor spiral is the S0 type of galaxy (**above**). S0s are more elongated than E7s, but do not show any spiral structure.

About one-third of spiral galaxies have huge bars running down the middle (**right**). Spiral arms trail from these bars rather like water from the spinning arms of a lawn sprinkler. These galaxies are called barred spirals, but some astronomers believe that almost all spiral galaxies, including our own Milky Way, contain at least the vestiges of a bar.

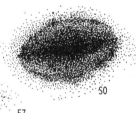

SBa

Just as with spiral galaxies, there is a three-stage classification for barred spirals from SBa types, which have large nuclei and tightly wound

Elliptical galaxies are old structures that formed within a billion or so years of the Big Bang. They consist entirely of old stars, typically red ones, and unlike spiral galaxies contain little interstellar matter. Elliptical galaxies are classified by shape. They vary from E0 types, which are almost spherical, like the galaxy M87 (**above**), through to E7 types, which have nearly flat disks, like spiral galaxies.

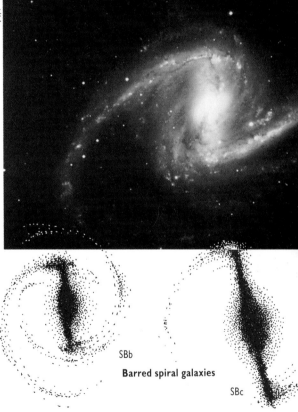

SBb

Barred spiral galaxies

SBc

arms, to SBc types, which have small nuclei and open arms. NGC 1365 (**above**) in the Fornax cluster of galaxies is an SBb type galaxy with a well-defined bar of stars across its center.

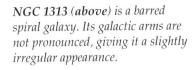

NGC 1313 (*above*) *is a barred spiral galaxy. Its galactic arms are not pronounced, giving it a slightly irregular appearance.*

Three-quarters *of all galaxies are round – either elliptical or spiral – but the rest seem to have no obvious shape at all, such as the Large and Small Magellanic clouds. These are known as irregular galaxies. NGC 6822 (*below*) is an irregular found in our local cluster of galaxies. Some irregular galaxies appear to have an underlying symmetry akin to that of spiral galaxies. But they may contain fewer than a billion stars, and astronomers think that is not enough for them to develop a spiral structure.*

ARTHUR STANLEY EDDINGTON – GALACTIC PIONEER

Among the most eminent of all 20th-century British scientists, Sir Arthur Stanley Eddington (1882–1944) was an expert in astrophysics. In his paper, *Stellar Movements and the Structure of the Universe*, published in 1914, he put forward the idea that spiral structure "nebulae" were in fact separate galaxies, similar to the Milky Way.

Eddington was also famous for establishing the link between a star's size and its brightness and for explaining the importance of the balance between the inward gravitational pressure and the outward pressures of radiation and gas in stars.

In 1919, he led an expedition of astronomers to Príncipe Island, off the west coast of Africa, where he was able to provide practical proof of Einstein's theory of general relativity. During a total eclipse, he measured a slight deflection of starlight caused by the gravitational effect of the presence of a large body – the Sun.

Eddington was the first to write about relativity in the English language, and his works on the subject were regarded as some of the finest by the theory's originator, Albert Einstein.

Sir Arthur's other published works on astrophysics include *The Internal Constitution of Stars* (1925) and *Stars and Atoms* (1927).

Galactic evolution

Like everything else in the universe, galaxies are dynamic entities, constantly changing and evolving.

No one is yet sure how galaxies were created. Some astronomers believe that they formed from vast concentrations of gas created by the Big Bang, which then broke up into galaxy-sized clouds – the "top-down" theory. However, the majority of observations seem to point toward a condensing of gassy material into clouds, first by collision, then under the pull of their own gravity. This is the "bottom-up" theory and may explain why there are more small galaxies than large ones. It may also explain why very few young galaxies exist – the early galaxies all formed relatively quickly from the available gas – and why galaxies are still merging to form clusters and superclusters.

Condensation of the primordial gas clouds led, it seems, to compression and the eventual ignition of a first generation of stars, forming a vast halo around a central core. In the core, new stars formed so rapidly that infant galaxies burned brilliantly – they were perhaps a hundred times brighter than the galaxies of today. Around these ancient cores were to form the structures of the galaxies we now see in our night sky.

The formation of spiral arms has puzzled astronomers. One theory explains them as being due to gravitational disturbances in the galaxy's disk which cause density waves – moving concentrations of matter. Something similar happens when fast-moving traffic on a two-lane highway meets an obstruction and is forced through a bottleneck by a slow-moving truck, before being able to pass it. At night, an aerial observer would see an even spread of lights before and after the bottleneck. The bottleneck itself would appear brighter because of the amount of traffic clustered around it. Cars approaching the truck would be forced to slow down until they could pass. As a result, the observer would see a slow-moving concentration of lights, the contents of which are, in fact, continuously changing as more traffic joins and leaves the bottleneck.

In much the same way, a density wave moving around a galactic disk causes large clouds of hydrogen gas to concentrate until there is pressure enough to condense them further and cause star formation and a blaze of light. At first the group of new stars is not in a spiral shape, but the movements of stars around the center of the galaxy slowly make the group into a spiral arm, since stars farther out orbit more slowly than those closer to the center.

Fast–moving bodies approach obstruction

Slow–moving obstruction

Individual bodies are slowed down and bunched together while passing the slow–moving obstruction

Bodies ac beyond o

WHEN GALAXIES COLLIDE

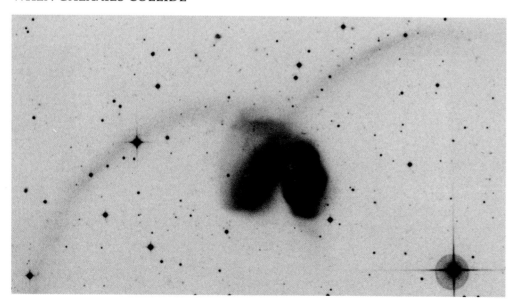

Structures such as the Antennae (left) – shown as a photographic negative – result from interactions between two galaxies, in this case NGC 4038 and NGC 4039. The galaxies are very close in space, and their mutual gravitational attraction has distorted them, pulling off tails of stars, gas, and dust over half a million light-years long.

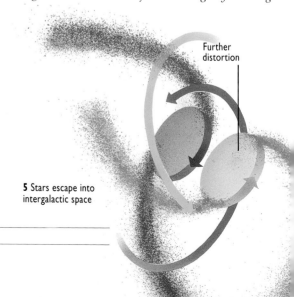

Further distortion

5 Stars escape into intergalactic space

Massive stars are concentrated in central region. Diffuse gas cloud begins to orbit.

Stars form slowly in outer regions. Partial disk develops.

Each part of a spiral galaxy forms at a specific stage of its evolution. Spiral galaxies consist of a nucleus, a disk, on which are found the spiral arms, and, in a spherical halo around the nucleus and arms, the globular clusters. The nucleus condenses from a huge cloud of spinning primordial gas which collapses under its own gravity. The oldest and more massive stars are found here and also in the scattering of globular clusters that are left behind once the cloud begins to condense.

Globular clusters

Central bulge

Spiral structure forms

Around the nucleus lie regions of diffuse gas in which the density is not enough to cause rapid star formation. As the galaxy rotates, the nucleus's gravitational pull, combined with the rotation, causes these gas regions to condense into a disk. Star birth continues slowly within the disk, and density waves cause spirals.

1 Approaching galaxy

2 Galaxies rotate in opposite directions

Galaxies are not alone in the universe, and the movement of one can affect another. In this hypothetical galactic encounter, two galaxies approach each other (**1**) rotating in opposite directions. When they meet, although in real terms they are still thousands of light-years apart, their mutual gravitational attraction sends them circling around each other (**2**),

Tidal forces distort disks

4 Streams of stars ejected

Disks become smaller

3 Galaxies begin to orbit common center of mass

Bridge of stars

*clinging together like two skaters. The immense combined mass involved means that huge tidal forces begin to act upon the galactic disks, distorting their shape as they orbit about their common center of mass (**3**). Gas clouds within are churned up, producing a blaze of new stars. As the galaxies continue to circle each other, vast trails of stars are ripped away, forming bridges and tails between the disks (**4**). This process occurs over many hundreds of millions of years and results in smaller disks with vast trailing tails, the stars of which may have gained enough velocity to escape into intergalactic space (**5**).*

Active galaxies

Echoes of the birth pangs of galaxies can come in the form of powerful emissions from active galaxies.

All galaxies are "active" to some extent, in that they give out radiation in the form of light, radio waves, X-rays, and so on. But some galaxies are very much more active than others. The most active are quasars, remote sources of energy thought by some to be galaxies early in their development. Less energetic and less distant types of active galaxy include Seyfert galaxies, which give out similar amounts of light to a galaxy like ours, but which send much more powerful radio signals out across the universe. Some active galaxies are, in fact, known as radio galaxies because they give out so much energy at radio wavelengths. It is thought that at the center of all types of active galaxies are black holes which power the energetic action.

Quasars (meaning quasi-stellar radio sources) must be extremely bright; otherwise, we would not be able to detect them at all because they are phenomenally far away. The light from quasars is shifted so massively to the red end of the spectrum that astronomers have judged many of them to be between 10 and 13.5 billion light-years away from Earth. This means the light we see from quasars started traveling toward

Lobes emit radio signals

us up to 13.5 billion years ago. The universe may be barely 15 billion years old, so when we look at quasars, we might be seeing infant galaxies, newly born to a youthful universe.

Active galaxies have been found closer to Earth, and they fill in the gap in brightness between our own galaxy and far distant quasars. Some of them have intensely bright, starlike cores that appear to be like mini-quasars. A Seyfert galaxy, for instance, is spiral shaped and, although it gives out little more visible light than an ordinary galaxy, it emits hundreds of times more infrared. However, its core can vary in brightness by millions of percent in the space of a few weeks.

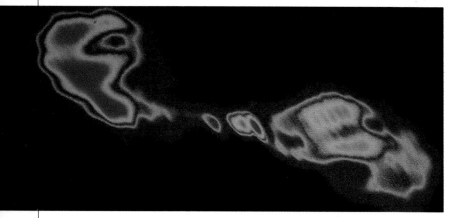

*Centaurus A is the nearest powerful radio galaxy, some 17 million light-years away. A radio image of the galaxy (**above**) clearly shows the double lobes of gas jets emitted from the nucleus. The structures shown here are referred to as the "inner lobes."*

*An X-ray image of Centaurus A (**above**) reveals a jet of matter, shown here in yellow, some 15,000 light-years long and emanating from the core of the galaxy. The emission is caused by synchrotron radiation that is produced by electrons in the magnetic field.*

A long exposure photograph of Centaurus A taken in visible light reveals the galaxy's elliptical structure. The active core emits strong X-ray and radio signals and lies at the heart of this structure, but is obscured by the gas and dust swirling around the galaxy.

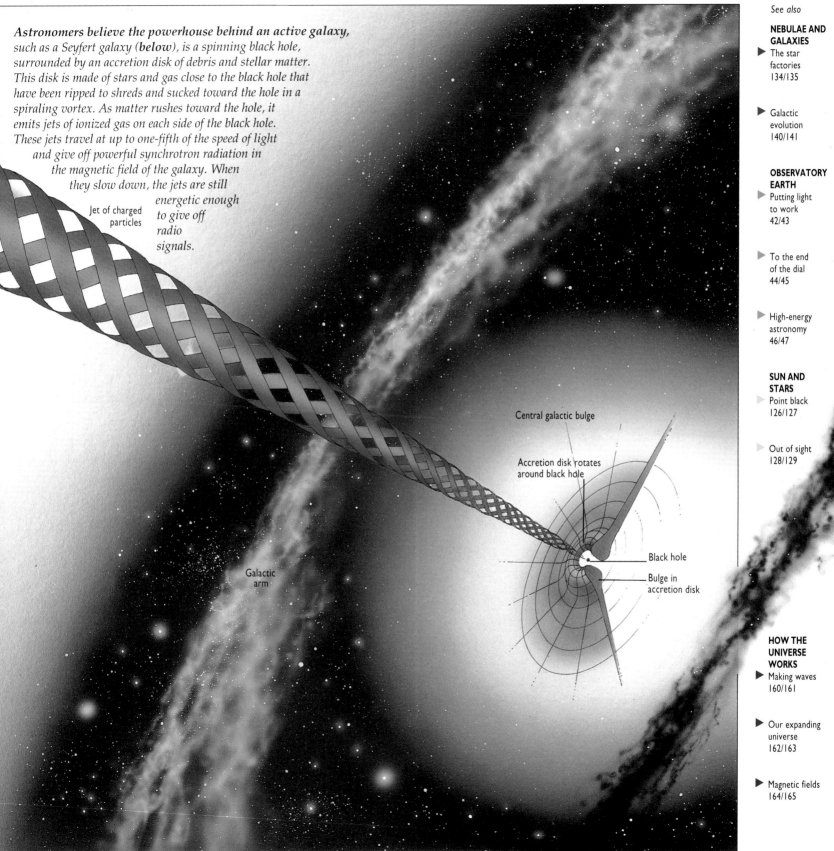

Astronomers believe the powerhouse behind an active galaxy,
*such as a Seyfert galaxy (**below**), is a spinning black hole,*
surrounded by an accretion disk of debris and stellar matter.
This disk is made of stars and gas close to the black hole that
have been ripped to shreds and sucked toward the hole in a
spiraling vortex. As matter rushes toward the hole, it
emits jets of ionized gas on each side of the black hole.
These jets travel at up to one-fifth of the speed of light
and give off powerful synchrotron radiation in
the magnetic field of the galaxy. When
they slow down, the jets are still
energetic enough
to give off
radio
signals.

Jet of charged
particles

Galactic
arm

Central galactic bulge

Accretion disk rotates
around black hole

Black hole

Bulge in
accretion disk

Our galaxy

The band of pearly luminescence stretching across the night sky is our own galaxy.

When you are right in the middle of something, it can be hard to get the big picture. Early astronomers had no idea that the Milky Way was, in fact, a galaxy seen sideways from within. Early explanations owed more to myth than science. This milky light was thought to be many things, including a path of the dead, the route to Zeus's palace, and the seam of the heavenly tent.

In 1917 American astronomer Harlow Shapley showed the true nature of the Milky Way. It is a spiral galaxy, a vast rotating disk-shaped gathering of stars, gas, and dust 100,000 light-years across and about 1,000 light-years thick. At its heart is a dense ball of old stars known as the central bulge. Around the bulge spins the disk of younger stars, concentrated into spiral arms. Our sun is in one of the spiral arms, some two-thirds of the way out from the central bulge.

Beyond the disk, a scattering of globular clusters and individual stars, plus what scientists believe to be dark matter, gives the Galaxy enough mass to stop it from flying apart.

Seen in infrared (above), the temperature variations in the Galaxy become apparent – warmest is blue, intermediate is green, and red is cool. The yellow-purple (warm) regions just above and below the center are areas of intense star formation in the constellations of Ophiuchus and Orion respectively. Distant galaxies appear as yellow-green dots spread across the sky.

The Milky Way is visible as a hazy band of bright stars across the night sky (below). Dark clouds are so thick at the center of the Galaxy that its bright heart cannot be detected at visible wavelengths.

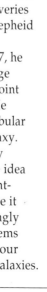

HARLOW SHAPLEY – SIZING THE GALAXY

Having started his professional life as a crime reporter, American farmer's son Harlow Shapley (1885–1972) switched to astronomy in his late 20s and made one of the most important astronomical discoveries of the century – the structure of our galaxy. Using stars called Cepheid variables as guides to distance, Shapley mapped out the three-dimensional distribution of 93 globular clusters in space. By 1917, he had shown that these clusters form a huge spherical system centered around a point in the Milky Way near Sagittarius. He then suggested, rightly, that the globular clusters show the extent of the Galaxy. In 1918 he published his Big Galaxy theory in which he put forward the idea that the Galaxy is some 300,000 light-years across, almost 10 times the size it was then thought to be. He also wrongly argued that the other spiral systems observed were associated with our galaxy and were not separate galaxies.

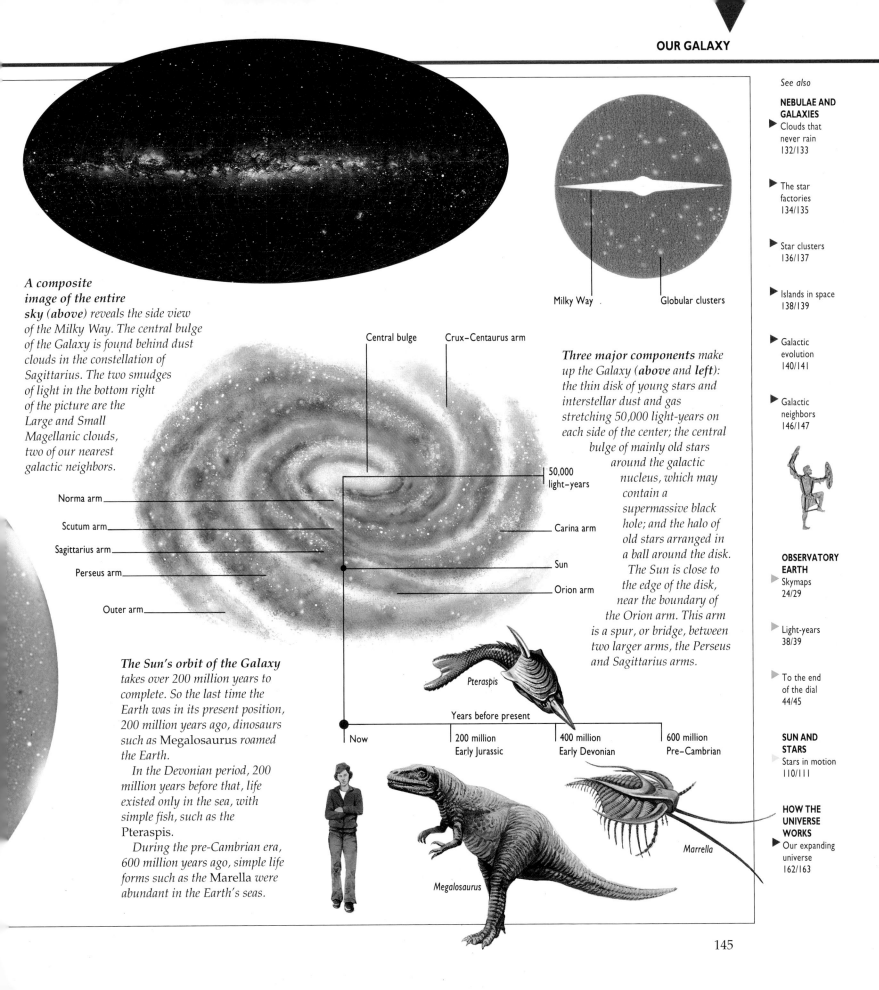

*A composite
image of the entire
sky (above) reveals the side view
of the Milky Way. The central bulge
of the Galaxy is found behind dust
clouds in the constellation of
Sagittarius. The two smudges
of light in the bottom right
of the picture are the
Large and Small
Magellanic clouds,
two of our nearest
galactic neighbors.*

Milky Way   Globular clusters

Central bulge   Crux–Centaurus arm

Norma arm

Scutum arm

Sagittarius arm

Perseus arm

Outer arm

50,000
light-years

Carina arm

Sun

Orion arm

Three major components make
up the Galaxy (**above** and **left**):
the thin disk of young stars and
interstellar dust and gas
stretching 50,000 light-years on
each side of the center; the central
bulge of mainly old stars
around the galactic
nucleus, which may
contain a
supermassive black
hole; and the halo of
old stars arranged in
a ball around the disk.
   The Sun is close to
the edge of the disk,
near the boundary of
the Orion arm. This arm
is a spur, or bridge, between
two larger arms, the Perseus
and Sagittarius arms.

The Sun's orbit of the Galaxy
*takes over 200 million years to
complete. So the last time the
Earth was in its present position,
200 million years ago, dinosaurs
such as* Megalosaurus *roamed
the Earth.
   In the Devonian period, 200
million years before that, life
existed only in the sea, with
simple fish, such as the
Pteraspis.
   During the pre-Cambrian era,
600 million years ago, simple life
forms such as the* Marella *were
abundant in the Earth's seas.*

Pteraspis

Years before present

Now

200 million
Early Jurassic

400 million
Early Devonian

600 million
Pre–Cambrian

Marrella

Megalosaurus

145

Galactic neighbors

*Although it seems vast to us, our galaxy, the Milky Way, is just
part of a clump of galaxies known as the Local Group.*

Throughout the night sky, it is possible to
see galaxies that are seemingly regularly
spaced in any given direction. This suggests
that the universe is the same everywhere, an
important idea called the cosmological
principle. Galaxies are not, however,
uniformly spread throughout the universe.
Instead, they show a tendency to clump
together into much larger structures, called
clusters. These clusters can include just half a
dozen or many thousands of galaxies. Our
own galaxy is part of a small cluster known
as the Local Group, which is nearly 7 million
light-years across.

There are, in fact, over 30 galaxies in the
Local Group besides our own. They range
from small irregular galaxies, such as the
Magellanic clouds, to large spiral galaxies,
such as the Milky Way and M31, the galaxy in
Andromeda. The cluster has no central focus;
instead, it is subdivided into two groups
centered on our galaxy and M31. Most of the
other galaxies, including the Large and Small
Magellanic clouds, are gathered around these
two large spiral galaxies. Our galaxy is, in
fact, orbited by 11 satellite galaxies, one of
which is the newly discovered Sagittarius
dwarf spheroidal galaxy. There is a third
large spiral galaxy, M33, in the Local Group,
but it is entirely alone.

There may be more galaxies in the Local
Group, but our view of them would be
blocked by a narrow zone which corresponds
with the plane of our galaxy. Here the large
amounts of gas and dust found within the
Milky Way absorb most of the radiation
coming from behind, making observation
almost impossible.

In the same way that planets move around
the Sun and stars move around the Galaxy,
the galaxies in the Local Group slowly orbit
around one another under the influence of
their combined gravitational attractions. The
orbits bring M31 toward us at 186 miles/s
(300 km/s) and send the Large Magellanic
Cloud away at 168 miles/s (270 km/s). Our
own Milky Way seems to be moving through
the group at about 78 miles/s (125 km/s).

When added together, the motions of all the
galaxies in the Local Group suggest that it
consists of much more matter than is visible.
Astronomers have theorized that it must
therefore contain large amounts of
invisible dark matter which,
when added to the
gravitational force of
the material that
is visible, holds
it all together.

*The Local Group (right) is shown centered on our own galaxy, the
Milky Way. The group contains over 30 galaxies, of which there are 3
giant spirals, 15 ellipticals, and 13 irregulars such as the Large and
Small Magellanic clouds. From this three-dimensional chart of the
cluster, it is easy to see that the group is focused upon the two larger
spiral galaxies, the Milky Way and the Andromeda galaxy (M31).
These two immense groups, along with several individual minor
galaxies, orbit around each other within a region with a radius of
some 3.5 million light-years. Across the vast reaches of space,
distance equals time, since light travels at a finite speed.
Thus light from the edge of the Local Group – 3.5
million light-years away – will have been
emitted 3.5 million years ago. At this
time, primitive human ancestors
were walking upon the
Earth.*

Key to types of galaxies

o   Elliptical

□   Irregular

🌀 Spiral

Visible to the naked eye, the Large Magellanic Cloud (**left**) is the nearest galaxy to the Milky Way. Seen in the skies above the southern hemisphere, its diameter is 16,300 light-years, about one-sixth that of the Milky Way.

Although it is 170,000 light-years away, the Large Magellanic Cloud is joined to our own galaxy by a stream of hydrogen gas clouds. The hydrogen was torn away from the Magellanic Cloud by the gravitational forces experienced between the two galaxies when they passed close to each other.

Mammuthus primigenius (*right*) existed 0.5 million years ago when light left this point.

Homo erectus (*right*) existed some 1.5 million years ago when light left this point.

Thylacosmilus (*right*) existed 2.5 million years ago when light left this point.

Light from the edges of the Local Group will take 3.5 million years to reach the Milky Way. We see objects at this distance as they were when the first hominid, **Australopithecus afarensis** (*right*), appeared on Earth.

22h 21h 20h 19h 18h 17h 16h 15h 14h 13h 12h 11h 10h 09h

DDO 210
Draco Ursa Minor
Milky Way
NGC 6822
Leo II
Leo I
Leo
Sextans I
Sculptor
Fornax
IC 5152
Carina
Large Magellanic Cloud
Small Magellanic Cloud
Sagittarius
Leo A
GR8

0.5 1.5 2.5 3.5

Millions of light-years

147

Clusters and superclusters

Seen in its entirety, the universe seems to be like a collection of massive soap bubbles held together by invisible strings.

Beyond our own Local Group of galaxies are many millions of clusters and superclusters of galaxies. They are all so far away, however, that only the brightest concentrations are visible from Earth. One nearby cluster is the vast Virgo cluster, which is over 50 million light-years away and contains more than 1,000 galaxies. The Virgo cluster is so big that three of the giant elliptical galaxies within it are each 2 million light-years across.

Clusters are classified as rich and poor, and further divided into regular and irregular. Virgo is rich, because it has so many galaxies within it, and irregular, because of the random scattering of its galaxies. Regular clusters, such as the Coma cluster some 300 million light-years away, are gathered into a ball shape.

Many clusters associate with other clusters in huge groups known as superclusters, which contain millions of galaxies. The Local Group lies on the fringes of the Local Supercluster. The Hercules and Perseus superclusters are among others that have been found. They are separated from us by huge voids, which the galaxies surround like the film around soap bubbles.

The view (right) shows the distribution of galaxies in the sky of the northern hemisphere. It highlights the tendency of galaxies to collect into huge superclusters, leaving large regions as empty voids. The image has been broken up into a million pixels, each shaded according to the density of galaxies found in that arc of space – from 0 (black squares) to 10 or more (white squares). The blank area at the top left is the southern horizon.

Because light travels at a fixed speed, objects that are a vast distance away appear to us not as they are now, but as they were when the light left them. The chart (below) compares the size of certain structures to events on Earth. Light would require 100,000 years to travel across the Milky Way, so objects at the far end of the Galaxy appear as they did during the ice ages of the late Pleistocene era. The Local Group, our cluster of galaxies, is nearly 7 million light-years across. Objects here appear as they were when the first bipedal apes existed.

Beyond this, the Local Supercluster is some 100 million light-years across. Light from the far end would have begun its journey some 100 million years ago, during the reign of the dinosaurs. Our supercluster is in turn dwarfed by the immense voids around it, measuring about 360 million light-years across. Light from the far end of these would have left during the early Carboniferous period.

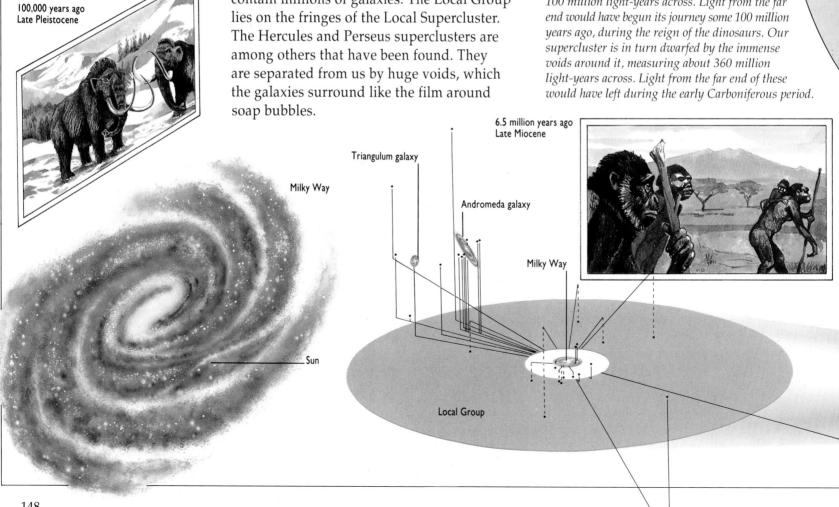

100,000 years ago
Late Pleistocene

Milky Way

Sun

Triangulum galaxy

Andromeda galaxy

6.5 million years ago
Late Miocene

Milky Way

Local Group

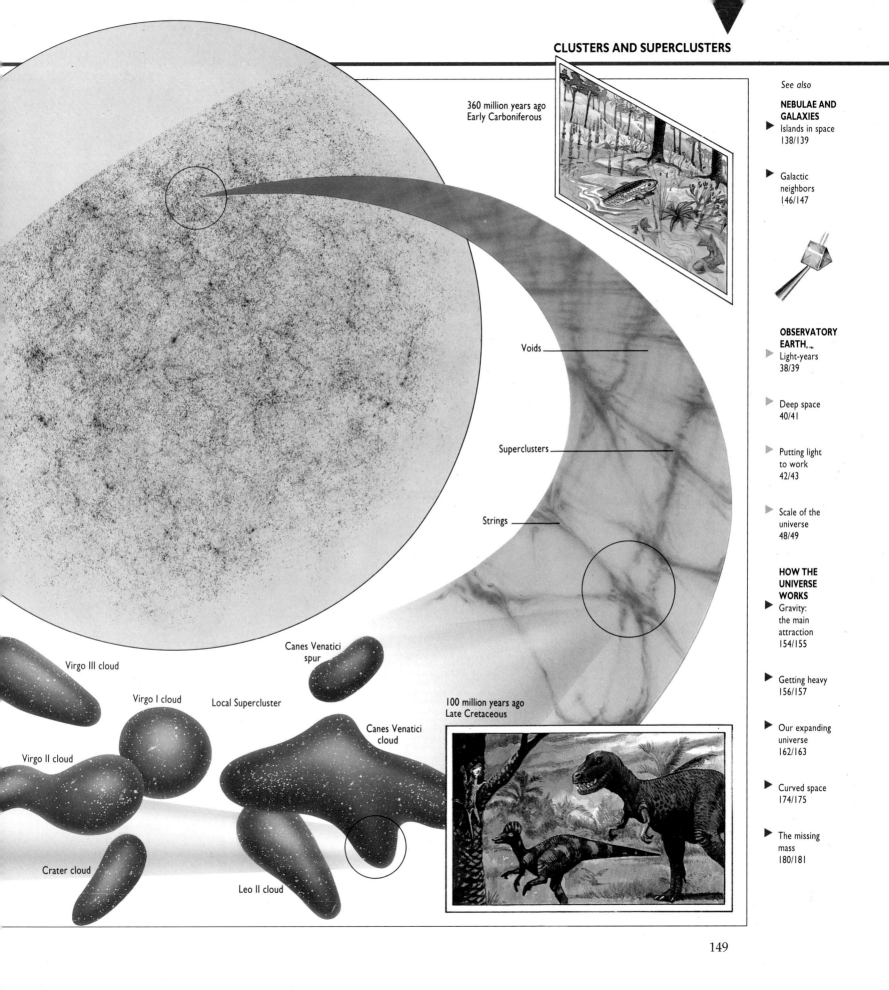

360 million years ago
Early Carboniferous

Voids

Superclusters

Strings

Canes Venatici
spur

Virgo III cloud

Virgo I cloud

Local Supercluster

Canes Venatici
cloud

Virgo II cloud

100 million years ago
Late Cretaceous

Crater cloud

Leo II cloud

See also

**NEBULAE AND
GALAXIES**
▶ Islands in space
138/139

▶ Galactic
neighbors
146/147

**OBSERVATORY
EARTH**
▶ Light-years
38/39

▶ Deep space
40/41

▶ Putting light
to work
42/43

▶ Scale of the
universe
48/49

**HOW THE
UNIVERSE
WORKS**
▶ Gravity:
the main
attraction
154/155

▶ Getting heavy
156/157

▶ Our expanding
universe
162/163

▶ Curved space
174/175

▶ The missing
mass
180/181

149

How the Universe Works

F rom the everyday happenings in our world to the extraordinary events that can take place in space, just a few principles and rules govern the way the universe operates. For instance, only four fundamental forces control everything that takes place in the universe. And we find that light always travels at a fixed speed, and gravity, space, and time all affect each other. The apparent complexity of the universe can be unraveled with a knowledge of how the principles and rules operate, whether on the smallest subatomic scale or across the entire universe.

Using this knowledge, astronomers are now able to probe back to near the beginning of time and describe what went on during the most important moment this universe will ever know – the Big Bang. But perhaps most amazing of all, there is life in the universe that has the intelligence to contemplate and work out its grand design.

Left (clockwise from top): In a spin with magnetism; the wheel of light; the particles that matter; it's a gas – pumping up a tire; speed and wavelength.
This page (top): The elements of fossil dating using the weak nuclear force; (left) a matter of some gravity – the Earth's tides and the pull of the Moon.

Forces in the universe

Everything that happens throughout creation takes place as a result of just four fundamental forces.

An almost infinite range of events is going on in the universe, from the buzzing of a bee to the explosion of a giant star. And it might seem that, because nothing changes without force, the range of forces is as huge as the range of events. Yet all forces either push or pull – sometimes directly, like a tug on a rope, sometimes indirectly across space, like the attraction between a magnet and a piece of iron. Every single change in the universe can be reduced to the operation of just four forces.

Two of the four – gravity and the electromagnetic force – are familiar and work on a large scale. Gravity holds the entire universe together, and the electromagnetic force is responsible for light. The other two forces are the strong and the weak nuclear forces, which operate only within the nuclei of atoms.

Physicists are now beginning to link the four forces. The weak nuclear force and electromagnetism are thought to be components of a single, electroweak force. And possibly the strong nuclear force will also be incorporated to make an "electromagnetic nuclear force." Research continues, its aim to show that underlying all four forces is a single superforce.

*The electromagnetic force makes the filament in a desk lamp (**right**) shine brightly. When the light is switched on, an electric current, made up of a stream of electrons, flows through the filament. The electrons bombard the atoms within the filament and give them energy. This heats up the filament and makes the atoms so energetic that they give off the energy they have received as visible light.*

It is on a subatomic level that the electromagnetic force begins. Certain subatomic particles, including electrons and protons, are charged, which simply means they either attract or repel each other. Particles with unlike charges attract; those with like charges repel. An electron carries a negative charge and a proton carries a positive charge, so electrons are attracted and bound to protons in the nucleus of an atom by the difference in charge.

No one knows quite what electrical charge is, but it seems to involve the transfer of photons, minute particles which have no mass but travel at the speed of light. Photons help create a large area of electromagnetism or "field" around every charged particle or group of particles. It is photons, also, that carry energy in the form of visible light away from the heated filament. Photons are, in fact, the particles involved in the transfer of all types of electromagnetic energy.

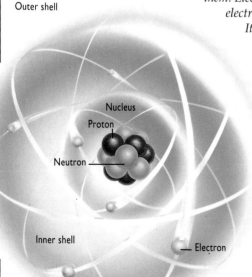

Outer shell

Nucleus

Proton

Neutron

Inner shell

Electron

*Three out of the four forces – electromagnetism and the strong and weak nuclear forces – hold atoms (**left**) and molecules together, or push them apart or change them. Electromagnetic attraction holds electrons to the nucleus of an atom. It also binds one atom to another, enabling molecules to be built. The strong nuclear force binds neutrons and protons together in the nucleus. The weak nuclear force allows an atom of one element to become that of another when it acts in the change of a neutron to a proton. Gravity, the fourth force, is so weak that its effects are irrelevant except on the largest scale, though each particle attracts every other.*

Every event, small or large, in an everyday scene (**left**) is caused by the operation of one or more of the four forces. Gravity keeps people – and the atmosphere and everything else – on the Earth. Electromagnetism provides the warmth of the Sun and the attraction between atoms that holds bodies and plants together. It also underlies the chemical reactions that enable humans to walk and plants to grow.

The strong nuclear force holds the nuclei of atoms together. It also drives the nuclear reactions that heat the Sun. Even the weak nuclear force is subtly present, helping to break up atoms slowly, changing, for instance, the chemical nature of the carbon in trees.

Gravity is the most familiar, yet least understood, of the four forces. It is the force that makes a dead leaf flutter down (**above right**) and that keeps your feet on the ground. It also holds the Earth and other planets in orbit around the Sun, prevents stars from disintegrating, and binds stars into galaxies.

Gravity is the force of attraction that exists between every kind of particle in the universe, pulling them all together. Gravity is, in fact, the dominant force in the universe, effective over huge distances. Like electromagnetism, gravity has an area of influence, or field. Although much weaker than electromagnetic fields, gravitational fields are vast in extent.

But despite holding the universe together, gravity is so incredibly weak that its effects are almost negligible on an atomic scale. Only on a scale the size of a planet or satellite or on a larger one, such as the size of a star, are enough particles concentrated close enough for their combined influence to make gravity a significant force in astronomy.

The weak nuclear force is the force that blasts the outer layers of an exploding supernova into space, creating a cloudy ring of expanding gas. But this is one of the few examples in which its effects are very obvious. Usually the weak force works subtly, as in the slow natural "radioactive" disintegration of certain unstable atoms. The steady breakup of carbon-14 atoms in dead organic matter by the weak force is like the ticking of a clock, allowing us to pinpoint the time of death by measuring the proportion of these atoms still intact, a technique called radiocarbon dating. Similarly, very old fossils (**right**) can be dated by measuring the amount of potassium-40 that remains in them.

Unlike the other three forces, which either pull or push, the weak force is a force for change. It comes into play when a neutron in the nucleus of an atom is converted into a proton, an electron, and a minute particle called a neutrino. When carbon-14 decays, for instance, one of the neutrons in the nucleus of the carbon atom changes into a proton, transmuting the element into nitrogen-14. The weak force operates over the smallest range of all the four forces: about the size of a proton or a neutron.

The strong nuclear force is the force that lies behind the most dramatic events in the universe – from the explosion of a nuclear bomb (**left**) to the burning of a star. All these events are nuclear reactions, and the strong force is the powerful force that binds the nucleus of each atom together. Because they all have like, positive electrical charges, without the strong nuclear force to hold them together, all the protons in the nucleus would repel each other violently and be hurled apart, and the neutrons would be held only loosely in place.

When atomic nuclei split (fission) or merge together (fusion) in nuclear reactions, the huge binding energy of the strong nuclear force is released. On Earth, fusion bombs (hydrogen bombs) make use of this to release a devastating amount of energy; in space fusion powers the stars.

Yet although the strong nuclear force is immensely powerful, the force field it creates is minute, barely reaching across the tiny nucleus of each atom.

See also

HOW THE UNIVERSE WORKS
▶ Gravity: the main attraction 154/155

▶ Getting heavy 156/157

▶ Making waves 160/161

▶ Magnetic fields 164/165 ·

▶ What is matter? 166/167

▶ Making the elements 168/169

▶ Forces in the atom 170/171

▶ In the beginning 178/179

OBSERVATORY EARTH
▶ Putting light to work 42/43

SUN AND STARS
▶ The nuclear powerhouse 106/107

NEBULAE AND GALAXIES
▶ Clusters and superclusters 148/149

153

Gravity: the main attraction

Each bit of matter in the universe exerts an attractive force on every other. The force is called gravity, and it keeps the universe together.

Every particle, whatever its size, exerts a gravitational force, but the strength of this force depends on the mass of the particle involved. The pull of gravity is, in fact, very weak – so weak that the gravitational attraction between individual atoms is negligible. Even a large building exerts a force that can be detected only with the most sensitive equipment. Yet when vast numbers of particles are combined in massive objects such as planets and stars, the force due to gravity can be intense.

The Sun, for instance, contains so much matter that its gravity holds the planets in tight orbits around it. The mutual force of many stars gathers them into galaxies and galaxies pull each other into clusters.

Gravity is a mutual attraction. A stone falling to the ground falls largely because Earth's gravity pulls it down, but the stone's gravity also attracts the Earth upward. The stone's mass compared to that of the Earth is so small that its gravitational effect is tiny, and the stone falls and the Earth does not seem to rise. But when objects are closer in size, the two-way attraction is more obvious.

The force of gravity distorts objects slightly, in a so-called tidal effect. Tidal effects are common throughout the universe; the most familiar one is that of ocean tides on Earth.

THE MOON AND THE TIDES

Gravity holds the Moon in orbit around the Earth, but just as the Earth attracts the Moon, so the Moon pulls on Earth. The Moon exerts a force pulling the Earth into an oval stretched toward the Moon. As the Earth is relatively rigid, it resists this tidal-effect stretching, distorting by little more than 8 inches (20 cm). Water in the oceans flows freely, however, so it moves toward the side of the Earth facing the Moon, attracted by the Moon's gravitational force. Water also bulges out on the side of the Earth facing away from the Moon since the gravitational force of the Moon is less on this side of the Earth because it is more distant from the Moon. High tides thus occur on opposite sides of Earth simultaneously.

Tides rise and fall twice a day when tidal bulges pass by. The tidal cycle lasts 24 hours 50 minutes, not 24 hours, because the Moon covers a twenty-eighth of its orbit during 24 hours and is thus above a given spot 50 minutes later each day.

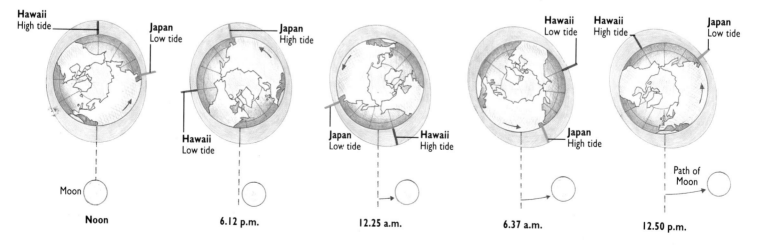

Day 28

Spring tide
New Moon

Day 21

Neap tide
Last quarter

Day 14

Spring tide
Full Moon

Day 7

Neap tide
First quarter

Sun's gravitational pull

Moon's gravitational pull

Day 0

Spring tide
New Moon

Moon

Direction of Sun

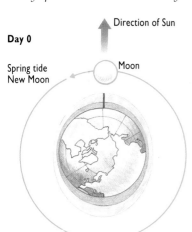

When a bungee jumper launches himself (above), Earth's gravity accelerates him downward so he plunges faster and faster until the elastic safety line brings him to a halt. Gravity affects all things equally, so all objects, whatever their mass, accelerate earthward at the same rate – at about 32 feet (9.8 m) a second each second. But as speed increases, air resists more. If the jump were to last long enough, the air resistance would match the force of gravity, and the jumper could fall no faster. He would then drop at a steady speed – his terminal velocity.

The gravitational forces of both the Sun and the Moon have a tidal effect on the Earth. Added together the two forces create a 28-day tidal cycle. Because the Sun is so far away, its influence is less marked than that of the Moon, even though the Sun is much more massive. The most dramatic effect occurs when Moon and Sun line up, and their gravitational pull is combined to produce extreme ocean tides on Earth called spring tides. At these times the highest high tides and the lowest low tides take place. Seven days later, when the Sun and Moon are at right angles to each other, however, their tidal effects counteract each other, producing less pronounced tides, called neap tides, with the lowest high tides and the highest low tides.

Another spring tide occurs 14 days into the cycle. At 21 days the neap tide repeats, and at day 28 another spring tide is experienced. The Moon takes 28 days to circle the Earth, so spring and neap tides each occur twice every 28 days.

Getting heavy

Everything from launching a satellite to estimating the density of a galaxy depends on understanding gravity.

The universe is full of orbits and movement: satellites orbit planets; planets orbit stars; stars move around the centers of galaxies; and galaxies revolve in clusters. They all turn this way because of gravity.

Celestial objects start moving through space for any number of reasons, but it is gravity that contains their motions. Without gravity, moving objects would hurtle off in straight lines and in all directions. But gravity pulls them into circular and elliptical orbits.

The shape and speed of the orbit depends on the strength of the gravitational force – the greater the pull, the higher the orbital speed and the tighter the orbit. The gravitational force, in turn, depends on the mass of the objects involved and the distance between them. The more massive they are, and the closer they are, the more powerfully they attract each other.

This simple relationship was discovered and put into a formula over 300 years ago when British mathematician Isaac Newton (1642–1727) set out his universal law of gravitation and the modern laws of motion. Obeying this simple law, massive objects close to the point around which they move orbit much faster than less massive, lighter ones farther away.

*Newton's law of gravitation shows that the attraction between two bodies increases with greater mass and decreases with the square of the distance between them, a relationship called the inverse square law. Thus the force between a planet and a satellite at **A** (below) is four times that between the planet and an identical satellite at **B**, only twice as far away.*

In the solar system this law is able to explain almost entirely the time it takes for planets to orbit the Sun. For instance, Mars, which has only about one-and-a-half times the distance to travel in its orbit compared with Earth, actually takes twice as long to complete that orbit. This is because its orbital velocity is lower since the Sun's gravitational force is reduced with the greater distance. In an Earth year, Jupiter covers barely 8 percent of its orbit and Saturn just 3.3 percent. Venus, however, which is closer to the Sun, completes just over one-and-a-half orbits. Mercury, closest to the Sun, orbits over four times in an Earth year.

A craft like the shuttle (above) uses its rocket motors to boost it into orbit. An artificial satellite, such as the shuttle, "falls" around the planet when it orbits. Gravity pulls the satellite down and this, combined with its forward speed, makes it move in a curved path. If its forward speed is high enough, the curve of the satellite's path matches the curve of the surface of the Earth it moves over. When this happens, the satellite orbits at a constant height above the surface.

Planet

Gravitational attraction

A

Satellite

B

Satellite

Inter–planetary probe launched with sufficient velocity to escape Earth's gravity

Low–velocity weather rocket falls back to Earth

The relationship between speed, mass, and distance is so precise that astronomers use Newton's law of gravitation to calculate the mass of a planet merely from the time it takes one of its satellites to orbit, and the satellite's distance from the planet.

The math becomes much more complex when more than two objects are involved. Nonetheless, the mass of a galaxy can be estimated simply by analyzing the motions of its stars. Knowing that the Sun lies about 30,000 light-years from the Galaxy's center and that it orbits at nearly 513,000 mph (825,000 km/h), astronomers have worked out that the mass of the visible part of the Galaxy is more than 100 billion times that of the Sun, or 100 billion solar masses. However, evidence shows that there is a lot of dark, invisible matter as well.

Satellite moving at orbital velocity "falls" in orbit around the Earth

To escape altogether from Earth's gravitational field, an interplanetary space probe must be accelerated to a speed known as escape velocity, which is about 7 miles/s (11.2 km/s).

To orbit the Earth, a probe has to be boosted above the slowing effects of the Earth's atmosphere. The minimum speed necessary to keep a craft orbiting around the Earth safely above the planet's atmosphere is 17,700 mph (28,500 km/h).

When a satellite is launched, the rockets must stop firing at exactly the right moment to set the satellite moving at the correct speed on its curved orbit. In order to reach this kind of speed, a spacecraft needs a boost from huge liquid-fuel filled rockets which are jettisoned when the craft is traveling fast enough.

Any launch below the minimum speed does not get the craft into orbit, and gravity pulls it in a loop back down to Earth, a fact made use of in weather probe rocket launches.

THE SEARCH FOR GRAVITONS

Gravitational force seems to travel across empty space instantly, without any obvious means of doing so. In recent years, however, physicists have begun to speculate that gravity might travel in waves made of particles called gravitons, just as light travels in waves of photons. Gravitons, like photons of light, would act as messengers carrying the "gravity signal" between objects in gravity waves. It is thought that these gravitons speed back and forward between the Earth and the Sun, for instance, keeping our planet in orbit. But no one is quite certain of their existence. If they do exist, such particles would be difficult to detect, even with a sophisticated machine like this gravity wave detector (**right**).

The big squeeze

The simple relationship between gravity, pressure, and temperature has huge implications in the universe.

Wherever matter accumulates, it is constantly squeezed together by the mutual gravitational attraction between particles. Even in deep space, where particles are widely scattered, this squeezing can raise temperatures considerably, making gases glow. Inside stars, the gravitational attraction of much matter can squeeze particles together so intensely that atoms fuse or even degenerate into their component parts.

Particles of matter are never still; they are whizzing about all the time. The hotter matter gets, the faster the particles move. There is rarely any overall direction to this movement, so particles frequently collide. Each particle, no matter how tiny, has mass. And when a moving mass hits something, it exerts a force. The constant bombardment of particles by other particles, and thus the constant exertion of myriads of tiny forces, creates the pressure in gases.

Gas pressure is highest when a gas is dense and there are more particles to bump into each other, or when it is hot and they crash together more vigorously. In fact, pressure is always in direct proportion to density and temperature. This relationship, known as Boyle's law, holds in most situations where matter is in its gaseous state. The exceptions are stars, where pressure of their radiation upsets the balance, and the hearts of collapsed stars.

Without gas pressure, stars would not be born. The rise in temperature brought on by the

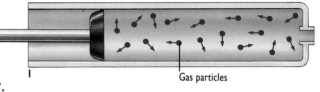

Gas particles

*Pressure and temperature changes occur when a tire is pumped up (**above** and **below**). Before the pump is pushed in (**1**), air in it is at the same pressure as air outside. When the pump is pushed in (**2**), its volume reduces and the air particles squeeze into a smaller space. As they jostle about, more hit the pump's sides and the force per unit area of the impacts – the pressure – rises. The particles also move faster since they pick up energy from the pump. This raises the temperature – heating the pump – and the pressure. When pressure in the pump is higher than that of the air in the tire, air is forced into the tire, inflating it.*

Diamonds are made of carbon – the same element as coal. But pressure makes them different. Although minute compared with the pressure inside the densest stars such as neutron stars, extreme pressures in Earth's crust can be enough to transform matter. The pressure built up in the crust caused by the heat of an erupting volcano can be quite adequate to rearrange the rather loose bonds between carbon atoms in a carbon molecule and force them into the incredibly strong lattice arrangement that makes diamonds so hard.

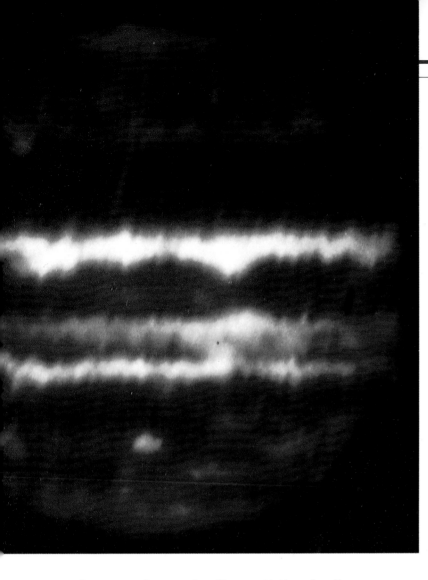

Pressure can make planets hot. *In 1969, Frank Lowe of Arizona University discovered that the planet Jupiter appeared to have an internal source of heat. The amount of heat coming from Jupiter has been estimated as being equal to the warmth from 4,000 trillion 100-watt light bulbs. All this energy is thought to result from the shrinking of the planet by barely ⅟₂₅ inch (1 mm) a year under the influence of its own gravity. This shrinking squeezes the particles of the planet's atmosphere together and raises their energy levels. In turn, this raises their temperature, and in the planet's interior the temperature reaches a blistering 30,000K.*

*The heat gradually moves upward to the top of the atmosphere where it is radiated away. This radiation has been detected – the brighter bands in this infrared (heat radiation) image of the planet (***left***) show where the heat from below is being emitted into space.*

In visible light, Jupiter (above) merely shines *by reflecting sunlight. But it seems that Jupiter gives out about two-and-a-half times as much heat as it receives from the Sun, thanks to the rise in temperature in its interior because of increased pressure.*

Extreme pressure, like that found in the central regions of Jupiter, does not only cause the temperature to rise, however. When a gas is squeezed really hard, its particles are pressed together so tightly that they change state from being gaseous to being liquid. Under even more extreme pressure, even the liquid changes, becoming solid.

pressure increase of a gas cloud's gravitational collapse eventually leads to conditions in which fusion can start and a star can radiate energy. Small and medium stars burn for billions of years because the pressure of gas within them counteracts their tendency to collapse under the influence of their own gravity. As gravity pulls inward, squeezing particles together, so nuclear reactions raise the temperature of the star's interior and boost the pressure, pushing the particles apart. Stable stars are in equilibrium, with gravity exactly balanced by internal pressure. Once the star's nuclear fuel finally burns out, however, pressure drops and gravity rapidly crushes the star. When this happens, pressure can rise again until matter breaks down into neutrons.

Pressure can also transform matter outside stars. In the depths of a gas giant planet's atmosphere, for instance, gas is under such pressure that it becomes a liquid. Once matter is liquid or solid, however, Boyle's law no longer applies because, generally, matter in these states is incompressible.

Making waves

Understanding light is vital to making sense of the universe.

Light and other forms of electromagnetic radiation provide almost our sole link with the rest of the universe. We know about distant stars, galaxies, and nebulae only as a result of the radiation they emit.

Most emission of electromagnetic radiation is intimately linked to the structure of the atom: it is given out whenever an electron changes its speed or direction. Typically, this occurs when an electron drops into an orbit closer to the nucleus of an atom and in the process throws off energy in the form of electromagnetic radiation.

Normally, atoms are in a "ground" state, with electrons orbiting close to the nucleus in the lowest possible energy level. So to give out light, an atom must first be "excited" – that is, it must absorb energy to boost an electron to a higher energy level and thus an orbit farther from the nucleus. This energy can come from electromagnetic radiation – in the form of photons – given out by other atoms.

Because electrons are restricted to certain sizes of orbits, they do not fall back inward gradually, but jump directly from one orbit to another, sometimes taking just one step, sometimes leaping several at a time. For every jump inward, an electron emits an amount of electromagnetic radiation – a photon. Since an electron has a certain level of energy in each possible orbit, the energy lost – and so the energy and wavelength of the photon – depends precisely on the size of the jump.

Each type of atom has its own set of alternative energy levels, and thus its own jump or emission pattern. When the light from a glowing gas is passed through a slit

The multicolored lights of a fairground ride (above) depend on countless billions of photons being given out by countless billions of atoms. For every tiny spot of light that reaches the eye, an atom in the wheel has absorbed energy in some form, then re-emitted it as visible light.

and then split into a spectrum, it gives its own unique pattern of bright lines. Every element in its gaseous state – sodium, carbon, hydrogen, and so on – can therefore be identified by its distinctive color signature of bright spectral lines.

Just as atoms can only emit certain wavelengths, so they can only absorb certain wavelengths. Thus, in addition to an emission spectrum, each gas has an absorption spectrum with a unique pattern of dark lines, which can also be used to help identify the element or elements in the gas.

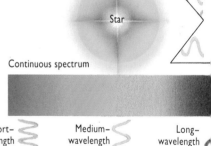

Continuous spectrum

Short-wavelength photon

Medium-wavelength photon

Long-wavelength photon

Star

The spectrum of an incandescent body, such as a star, is continuous (above), since it contains light of all wavelengths and a full range of colors. The wavelength of a photon determines how much energy it has: a photon at a short wavelength has more energy than a photon at a longer wavelength.

SYNCHROTRON RADIATION

Photons (the "particles" of electromagnetic radiation) are usually given off when the electron of an excited atom drops back to a lower energy level. However, under certain circumstances, electrons emit photons without an atom being excited at all. When an electron crosses a magnetic field, for example, the field accelerates it in a spiral following the lines of magnetic force.

If the electron is moving extremely fast, the acceleration stimulates it to emit bursts of radiation. The effect was first seen in a machine called a synchrotron (hence synchrotron radiation) at Daresbury Laboratory in Cheshire, England, but it also occurs naturally. Radio waves from quasars are synchrotron radiation. So, too, is radio radiation from Jupiter, which seems to come from vast numbers of charged particles spiraling around in its powerful magnetic field.

Electron path

Synchrotron radiation

Magnetic fields

For an atom to absorb a photon's energy, the photon must have exactly the right amount of energy (and thus a specific wavelength) to bump an electron orbiting the nucleus up to one of the atom's higher energy levels. In the gas cloud (**below**), the atoms absorb only the long-wavelength photons.

Because the gas has absorbed these photons from the star's light, there is now a gap in the spectrum of the star (**right**) that shows as a dark line.

Gas cloud

Light rays

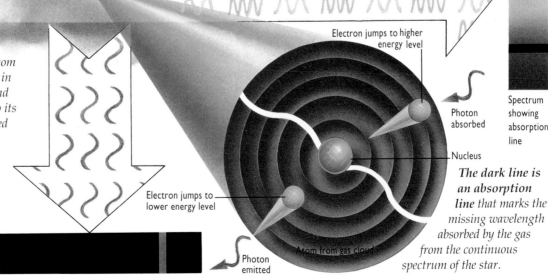

On absorbing a photon, the atom becomes "excited." It only stays in this state for a fraction of a second before the electron jumps back to its original orbit. The photon emitted in this jump carries a set amount of energy and is at a fixed wavelength. A spectrum taken of the cloud alone thus shows a bright emission line.

Spectrum showing emission line

Electron jumps to lower energy level

Photon emitted

Atom from gas cloud

Electron jumps to higher energy level

Photon absorbed

Nucleus

Spectrum showing absorption line

The dark line is an absorption line that marks the missing wavelength absorbed by the gas from the continuous spectrum of the star.

Our expanding universe

The color of light from distant galaxies shows that the universe is gigantic and getting bigger all the time.

Until the 1920s, it was thought that the universe consisted of little more than our galaxy – the Milky Way. The other galaxies seen in the heavens were reckoned to be clouds of gas. We now know that the Galaxy is just one of countless billions of galaxies in a universe so vast that most people cannot conceive how huge it is – if the Milky Way were the size of a pinhead, the universe would be bigger than a football stadium.

The universe is also growing by the second. In every direction we look, galaxies are speeding away from us at an astonishing rate – at up to half a million times the maximum speed of Concorde (1,450 mph (2,330 km/h). We know this because the light from them is altered by their movement. The light from distant galaxies is, in fact, shifted noticeably toward the red end of the spectrum – this change is called red shift.

The idea of red shift was first suggested in 1842 by the Austrian physicist Christian Johann Doppler (1803–1853), which is why it

The shifting of light wavelengths that occurs when objects move either toward or away from an observer cannot easily be seen. But its sound equivalent can often be heard, since Doppler shifts affect sound waves as well as light waves. When a motorbike approaches fast (left), its engine noise is at a high pitch. The sound waves shorten since the bike moves closer between the start of each wave.

When the bike draws level (below), the sound is at its true pitch for an instant. When the bike speeds away (far right), the pitch drops because the waves lengthen – the bike moves farther away between the start of each wave. The drop in pitch is like the red shift of the light from receding galaxies.

Motorbike approaches, sound wavelength shortens, engine pitch gets higher

is sometimes called the Doppler shift. Doppler suggested that light waves from a receding source are stretched out, because each successive wave is emitted from a position slightly farther away than one preceding it. If the light waves are stretched out, their wavelength increases and the light is made to appear redder. Conversely, waves from an approaching source squeeze up,

When motorbike is neither approaching nor receding the sound wavelength stays constant, the engine pitch remains unchanged

The universe expands like a balloon being blown up (left). The distances between galaxies – represented by stars stuck onto the balloon – increase as the balloon is inflated. The universe is expanding uniformly and, if the whole balloon grows at the same rate, the farther the stars (galaxies) are from each other, the faster they move apart. As the balloon doubles in size, for instance, stars 1 inch (2.5 cm) away from each other move 2 inches (5 cm) apart, while those 6 inches (15 cm) away move 12 inches (30 cm), etc. Individual galaxies do not expand because they are held together by the mutual gravitational attraction of all the matter of which they are made.

shortening their wavelength and making them seem bluer, a phenomenon called blue shift. But French physicist Armande-Hippolyte-Louis Fizeau (1819–1896) pointed out that no total color difference would be detected. What would be seen was a shift in the positions of the lines of the spectrum. The amount of the shift depends on the source's speed – the higher the speed, the greater the shift. It is simple to calculate the speed by measuring the amount of shift in the spectrum of an object.

When the American astronomer Edwin Hubble analyzed the red shift of distant galaxies in the 1920s, he showed that the farther away galaxies are, the faster they are receding from us. Their recessional speed increases with distance

Motorbike recedes, sound wavelength lengthens, engine pitch gets lower

at a steady rate – the Hubble constant. The exact value of this constant is a source of debate among astronomers. Figures vary between 43 and 57 miles/s (69 and 91 km/s) per million parsecs. Many current estimates put it at about 47 miles/s (75 km/s) per million parsecs. This means that a galaxy 30 million parsecs (about 100 million light-years) distant is speeding away from us at over 1,400 miles/s (2,250 km/s).

See also

HOW THE UNIVERSE WORKS
▶ Gravity: the main attraction 154/155

▶ Making waves 160/161

▶ Echoes of the Bang 176/177

▶ In the beginning 178/179

▶ The missing mass 180/181

OBSERVATORY EARTH
▶ Deep space 40/41

▶ Putting light to work 42/43

▶ Scale of the universe 48/49

NEBULAE AND GALAXIES
▶ Our galaxy 144/145

▶ Galactic neighbors 146/147

▶ Clusters and superclusters 148/149

EDWIN HUBBLE – EXPANDING ASTRONOMICAL IDEAS

Born in Marshfield, Missouri, Edwin Powell Hubble (1899–1953) was an athlete and boxer who gave up a promising career in law to study astronomy. In 1914 he joined the Yerkes Observatory, Wisconsin, which has a 40-inch (102-cm) refracting telescope. In 1923 he went to work at Mount Wilson Observatory, up a mule track high above Pasadena, California, and soon made the most important astronomical discoveries of the 20th century.

With the Mount Wilson telescope, which has a 100-inch (2.5-m) mirror and which was then the most powerful in the world, he photographed the smudge of sky known as the Andromeda galaxy and identified, for the first time, individual stars within it, 2 million light-years away and well outside the Milky Way. Having shown that there were galaxies beyond our own, Hubble went on not only to identify dozens more, but also to measure their red shifts with fellow American Milton Humason. Using this data, he established that all the galaxies in the universe are receding from us – and that the farther they are away, the faster they are receding. He announced this relationship in 1929, and it is known as Hubble's law. His work laid the foundation for the concept of an expanding universe and provided a way to calculate its overall size.

Hubble also introduced a widely used system for classifying galaxies, dividing them into spirals, barred spirals, and ellipses. The far-seeing space telescope, Hubble, launched in 1990, is named after him.

Magnetic fields

Some stars and planets both generate their own electricity and make powerful magnetic fields.

Fleming's left-hand rule shows the respective directions of the force a magnetic field exerts on a magnet in a motor, the field itself, and the electric current. With the first and second fingers and thumb of your left hand at right angles to each other (**left**), the second finger is the current, the first the field, the thumb the force.

Magnetism, together with electricity (of which it is an intimate part), is one of nature's four fundamental forces, and its effects are felt throughout the universe. Many phenomena, from the powerful radio signals emitted by distant quasars to the aurorae that light up the Earth's polar skies, are caused by electromagnetism.

Magnetic fields are doughnut-shaped regions of concentric lines of force around every magnet. Any magnetic or charged particle entering the field is swiveled at once to align with these lines of force. Such particles – and even heavy magnetizable objects – that come close to a magnet are drawn powerfully toward one end, or pole, of the magnet and pushed equally powerfully away from the other.

When charged particles such as electrons enter an intense magnetic field, they may be accelerated around the lines of force so much that they fire out photons of electromagnetic radiation. This is synchrotron radiation, and it is in the form of radio signals. Such radio signals alerted astronomers to Jupiter's strong magnetic field, and also to the presence of distant, energetic quasars and active galaxies.

There are other effects of strong magnetic fields that can be detected. They split certain lines in the light spectrum coming from distant objects, a result known as the Zeeman effect, named after Pieter Zeeman, the Dutch physicist who discovered it in 1896.

Wherever magnetism occurs in the universe, it is created by an electric current, and all magnets have two poles – a north and a south. On a subatomic scale, magnetism is created by electric currents set up by electrons as each one orbits the nucleus. So every atom is a little magnet. In most cases, these tiny atomic magnets cancel out each other's

*Electric motors, like those in a fan (**right**), use the same effect that creates a stellar or planetary magnetic field. Flowing current (here, around a coil of wire) sets up a magnetic field. The coil sits inside a magnet, and the combined effects of the two fields cause the coil to spin.*

164

magnetism. In ferromagnetic materials, such as iron and steel, however, they can line up to make a strong magnet.

On a cosmic scale, magnetic fields around stars and planets are created by electric currents generated as the body's liquid core swirls about. To have a strong magnetic field, the core must be electrically conducting – that is, metallic. The body must also spin rapidly enough to set up eddies in the core. The Moon, with its slow rotation and solid core, has no magnetic field. Venus has a liquid metal core but turns slowly, so its magnetic field is some 20,000 times weaker than Earth's. The rapid rotations of Saturn and Jupiter, along with their huge, liquid outer cores of metallic hydrogen, give them giant magnetic fields 20 to 30 times stronger than Earth's.

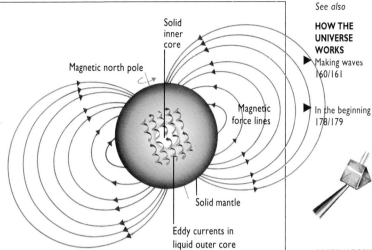

The Earth's magnetic field is created by slow eddies in its molten iron outer core (**above**). These set up electric currents which generate a magnetic field. This field, or magnetosphere (**below**), is bounded by the magnetopause. On the sunward side, there is a bow shock where solar wind particles hit the magnetic field, slow down, and compress the magnetosphere. The wind blows the magnetosphere out into a long tail on the opposite side.

Without this field to protect us, we would be bombarded by the solar wind – high-energy charged particles that stream from the Sun. The field traps these particles and spreads them along its lines of force in the Van Allen belt, where they spiral around generating radiation. Only at the polar horns can a few particles get down to the surface.

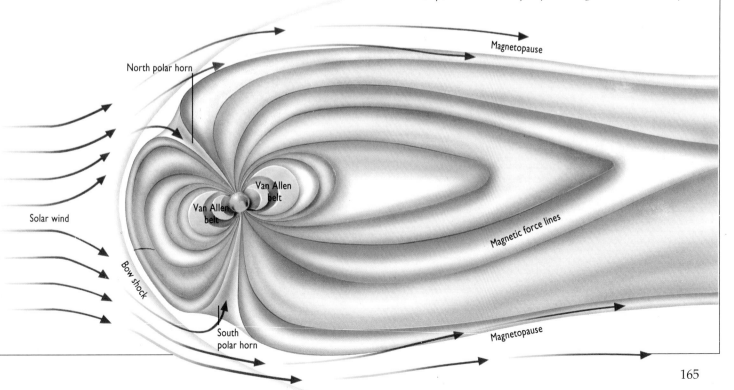

What is matter?

Although matter is made from a bewildering collection of particles, there may be an underlying simplicity and unity.

Before about 1920, just three kinds of subatomic particles were believed to be the basic building blocks of atoms and hence all substances. Atoms were thought to be made of neutrons and positively charged protons in a nucleus, with negatively charged electrons spinning around the outside. Since then, hundreds of other types of subatomic particles have been discovered, and many have mirror-image antiparticles. Most of these particles are short-lived.

One way to classify these particles is into hadrons, which feel the strong nuclear force, and leptons, which do not. There are six kinds of leptons, and the best known are neutrinos and the electron. There are many hundreds of different hadrons, including protons and neutrons, all closely associated with atomic nuclei. Hadrons are much bigger than leptons; a proton, for instance, is 1,836 times more massive than an electron. Most scientists believe that inside all hadrons are quarks, particles which come in six different "flavors" – up, down, strange, charmed, beauty, and truth. These join up in different flavor mixes: either in trios to make heavy "baryons" such as protons and neutrons; or in quark-antiquark pairs to make lighter but unstable "mesons."

Molecule
10^{-7} inch (10^{-9} m)

Atom
10^{-8} inch (10^{-10} m)

Nucleus
10^{-14} inch (10^{-16} m)

Matter is built up *from minute particles ranging from molecules to quarks. A small molecule like methane can be about 10^{-7} inch (10^{-9} m) across. Quarks are thought to be about 10 billion times smaller than a methane molecule at 10^{-18} inch (10^{-20} m) across. Methane is made of four hydrogen atoms and one carbon atom. Carbon has a nucleus that contains six protons and usually six neutrons, which is orbited by six electrons. Each neutron and proton in the nucleus is made of three quarks. The quarks' "flavors" determine whether a particle in the nucleus is a neutron or a proton.*

Hadron
10^{-15} inch (10^{-17} m)

Hydrogen atom

1 proton

1 electron

Quark
Less than 10^{-18} inch (10^{-20} m)

Hydrogen, the simplest and lightest element *in the universe (**above**), is also the most abundant. It has the smallest nucleus, with just one proton (and usually no neutron), and is completed by a single electron which orbits this nucleus. But, like all the other chemical elements, the number of protons in its nucleus determines its identity and characteristics. For each chemical element, there is a different number of protons in the atomic nucleus, up to a maximum of 109 – hydrogen has one, iron 26, gold 79, and so on. All other elements have nuclei that also contain neutrons. The more protons and neutrons an atom possesses, the heavier it is. All elements also have electrons, whose number is also determined by the number of protons in the nucleus. Protons and neutrons have approximately the same mass, but protons carry a positive charge and neutrons are neutral. Electrons are tiny fundamental, or indivisible, particles that are negatively charged and much less massive than either protons or neutrons.*

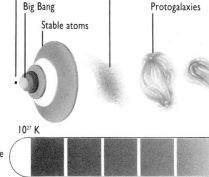

Singularity

Big Bang

Stable atoms

Expanding primordial gas

Protogalaxies

10^{27} K

Temperature

166

Of the matter that makes up the universe today (above), almost 90 percent is the element hydrogen. A further 9 percent is helium, the next lightest and simplest atom. All the remaining elements make up less than 1 percent of the total matter. The relative abundance of the different elements gives a clear hint to the order in which they formed. Most of the 107 elements we know today were synthesized as the universe evolved.

In the first moments of the universe, there was not a single chemical element, and for billions of years afterward there were no more than a handful. Hydrogen was the first element to form after the Big Bang, when the temperature cooled enough for protons to form from quarks. Soon after, some other simple atomic nuclei formed, including helium (right), which has just two protons and neutrons in its nucleus. After this initial period of atomic nuclei formation, no further elements emerged until fusion reactions began in the high-temperature cores of stars. Element formation in stars starts with the fusion of hydrogen to create helium, at a minimum temperature of about 10 million K, or 10^7 K, and then goes on to produce other elements.

MURRAY GELL-MANN

In 1963 an American theoretical physicist at Caltech (the California Institute of Technology), Murray Gell-Mann (1929–), together with George Zweig (1937–), provided the basis for an understanding of hadrons (such as protons and neutrons) by inventing quarks, particles that make up hadrons. (The word "quark" came from James Joyce's novel *Finnegans Wake*, in which one of the characters says, "Three quarks for Muster Mark!")

Gell-Mann (**below**) and Zweig originally suggested that quarks came in three "flavors" – up, down, and strange – but since the 1960s, three more quark flavors have been added – charmed, beauty, and truth.

With Richard Feynman (1918–88), one of modern physics' most colorful characters, Gell-Mann introduced the idea of currents, similar to electric currents, to explain the working of the weak nuclear force.

Helium atom

axies

Main sequence star

10^7 K

Conversion of hydrogen to helium

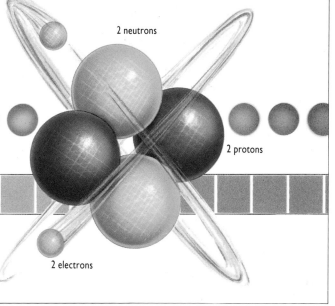

2 neutrons

2 protons

2 electrons

Making the elements

None of the chemical elements existed when the universe began – they have been built up, one by one.

Most of the elements we know were made by nuclear fusion in stars, starting from just hydrogen and helium. In order for nuclei to fuse, extremely high temperatures are needed, and the greater the mutually repulsive electric charges in the nucleus, the higher the temperatures must be. Since heavy elements whose nuclei have greater numbers of protons are generally more mutually repellent, it is thought that they were made in the hearts of massive stars, where the temperature is highest. The heavier the atom, the later in a star's life it is created, since only then are the pressure and temperature high enough to push the protons together against their repulsive forces.

THE MYSTERY PROCESS

Deuterium, lithium (**right**), beryllium, and boron are found in large quantities in cosmic rays, the jets of nuclei shot into space by supernova explosions at speeds close to the speed of light. It is not known quite how these elements are produced, so their synthesis is sometimes called the X-process.

Cosmic rays have provided scientists with much information about the variety of exotic particles – those other than protons, neutrons, and electrons. For instance, in 1932 Carl Anderson found the positron, the mirror image of the electron, in cosmic rays and so confirmed the existence of antimatter. And antiprotons were discovered in cosmic rays in the 1940s.

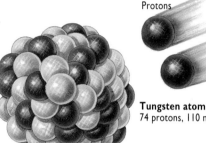

Protons

The P-process in the shock wave from a supernova can force protons onto light atoms in the hydrogen-rich shell of the exploding star to form rare heavy elements such as tungsten (**left**) and osmium.

Tungsten atom
74 protons, 110 neutrons

Proton

Silicon atom
14 protons, 14 neutrons

Carbon atom
6 protons, 6 neutrons

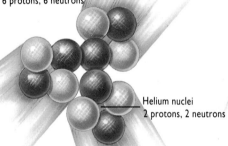

Helium nuclei
2 protons, 2 neutrons

*In the equilibrium or E-process, iron (**right**) forms when a massive supergiant nears the end of its life. Extremely high temperatures of over 1 billion (10^9) K turn silicon and other atoms into various types of iron, one of the most common of all elements. When iron fuses, it takes in energy, precipitating a supernova explosion.*

Iron atom
26 protons, 30 neutrons

Helium nucleus

Neutron

The capture of further alpha particles creates silicon (**above**), magnesium, argon, and calcium. These are made when pressure builds up so much that oxygen and neon nuclei begin to take on alpha particles. The addition of 3 alpha particles to oxygen's 8 protons and 8 neutrons makes silicon, which has 14 protons and 14 neutrons.

*Fusion of two helium nuclei (alpha particles) in red giant stars at a temperature of 100 million (10^8) K forms beryllium. Another alpha particle fuses on to make carbon (**above**), and a fourth one fuses on to make oxygen.*

Red giant

Supergiant

Exploding star

S-process

R-process

10^8 K

10^9 K

Temperature

├─ Capture of helium nuclei ─┤

Further capture of helium nuclei

P-process

E-process

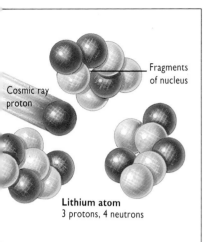

Cosmic ray proton

Fragments of nucleus

Lithium atom
3 protons, 4 neutrons

Around the heavy-element core of a dying star, helium burns slowly, releasing neutrons. Neutrons are not repulsed by the nucleus, so they can be added to nuclei in the so-called slow or S-process, which adds neutrons to iron and lighter isotopes to make bismuth (**right**), which has a total of 209 protons and neutrons. Bismuth then sheds an alpha particle to make lead.

When a giant star's internal resistance vanishes, it collapses rapidly and then rebounds in a huge supernova explosion. The shock wave rolling out from this explosion forces neutrons onto bismuth and iron nuclei in the R-process (rapid neutron capture) to make thorium, uranium (**right**), and similar large atoms. These and all the other atoms in the star are blasted into space to seed the gas clouds of the universe with heavy elements.

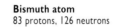

Neutrons

Bismuth atom
83 protons, 126 neutrons

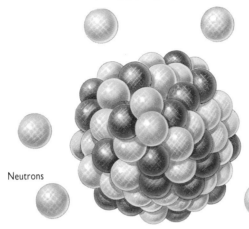

Neutrons

Uranium atom
92 protons, 146 neutrons

MOLECULES IN SPACE

The basic chemical units from which all substances are made are the atoms of the chemical elements such as hydrogen, carbon, oxygen, iron, and so on. These atoms can combine in a variety of ways to make molecular compounds – substances made of a fixed ratio of certain atoms. Water, among other molecules, is found in space, where its presence is detected by spectroscopy.

In water two hydrogen atoms bond with one oxygen atom, and in the resulting three-atom molecule the electrons are shared.

Water is, in fact, a remarkable compound. One of its properties is that it expands when it freezes, so that ice occupies about 9 percent more volume than liquid water. Ice cubes thus float in a glass of water.

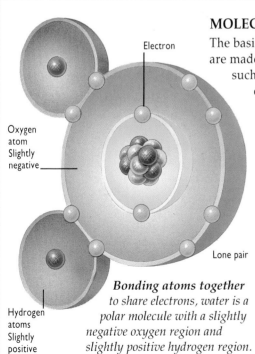

Electron

Oxygen atom Slightly negative

Hydrogen atoms Slightly positive

Lone pair

Bonding atoms together to share electrons, water is a polar molecule with a slightly negative oxygen region and slightly positive hydrogen region.

Forces in the atom

*The strong and weak nuclear forces bind particles
of atomic nuclei together – and break them apart.*

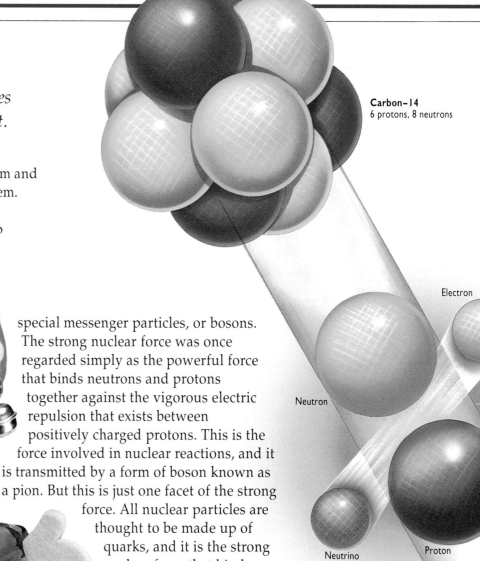

Carbon–14
6 protons, 8 neutrons

Electron

Neutron

Neutrino

Proton

Since the 1970s, physicists have made huge strides in understanding the nature of the forces inside the atom and have discovered hitherto unsuspected links between them. These links not only provide valuable insights into the processes going on inside stars, but may ultimately help to explain the workings of the universe as a whole. Crucial to these advances has been the development of quantum theory, which predicts that all four of the fundamental forces are carried by special messenger particles, or bosons. The strong nuclear force was once regarded simply as the powerful force that binds neutrons and protons together against the vigorous electric repulsion that exists between positively charged protons. This is the force involved in nuclear reactions, and it is transmitted by a form of boson known as a pion. But this is just one facet of the strong force. All nuclear particles are thought to be made up of quarks, and it is the strong nuclear force that binds quarks together, virtually unbreakably. It is as if quarks were held together like the links of a chain, which allows them to move apart so far, but no farther. Because of this extremely strong force, individual quarks are never seen.

The mass of the nucleus is slightly lower than the combined masses of the protons and neutrons that it is made of. But no mass is lost, because mass and energy are interchangeable. When protons and neutrons are thrust together in a nucleus, some of their mass turns into energy in the form of the strong force that binds them. This binding energy is trapped in the nucleus like a jack-in-the-box, waiting to spring. When atoms fuse (nuclear fusion) or split (nuclear fission), the lid comes off the box, and huge amounts of binding energy are unleashed. This is the basis of all nuclear energy.

Some atoms, especially large ones and those with an imbalance of protons and neutrons, are likely to break up under the influence of the weak force. A neutron that escapes from the nucleus of carbon-14 (above), for instance, is broken up by the weak force, creating a proton, an electron, and a neutrino. The proton returns to the nucleus, and the atom becomes nitrogen-14; the electron and neutrino are emitted, as is energy in the form of a photon.

Photon

Nitrogen–14
7 protons, 7 neutrons

Nuclear energy released by the strong force provides the only practicable source of energy for some long space flights. On board spacecraft – especially long-distance interplanetary probes such as the Viking landers which reached Mars in the mid-1970s (**left**) – are small nuclear reactors that obtain this energy from the fission of uranium nuclei, just as nuclear power stations do on Earth. Uranium contains a million times more energy per gram than oil – anything less concentrated would make the craft too heavy to launch.

The fuel that keeps stars burning for billions of years is also nuclear, but in stars it is the fusion (coming together) of small nuclei that releases the energy, not the fission (break up) of large ones. Deep inside stars, extreme pressure fuses hydrogen nuclei together to make helium nuclei, converting tiny amounts of the hydrogen nuclei's mass into enormous amounts of energy.

Neutrinos are among the most common particles in the universe, outnumbering electrons by a billion to one. Their existence was suspected when a slight loss of mass was noticed during beta decay – the break-up of a neutron to make a proton when an electron (beta particle) is given out. It was suggested that the missing mass was the then-unknown neutrino.

But neutrinos have proved very hard to detect because they have virtually no mass and are so little affected by gravity that they can pass clean through the Earth – and any detector – with no effect. Neutrinos respond to the weak force, but this operates over such a short distance – less than 10^{-16} mm – that the chances of a neutrino coming in range are small. Yet, neutrinos streaming from space have now been captured in fluid-filled detector tubes (**below**).

Although quarks cannot be separated, they can be altered, which is where the weak nuclear force comes in. The weak force does not, in fact, push or pull; it just changes things, and it causes radioactivity – the natural decay of particles. For instance, when a neutron decays into a proton, an electron, and a neutrino, the weak force changes one of the neutron's three quarks.

The weak force's bosons are massive and travel only a short distance before breaking up – which is why their effects are confined to individual particles. The weak force has three types of bosons: the W+, the W–, and the Z, each controlling a different force field. The W+ and W– are charged like protons and electrons; the Z is neutral like a neutron.

Everything is relative

No matter where you are or what you are doing, the speed of light is fixed – everything else is relative.

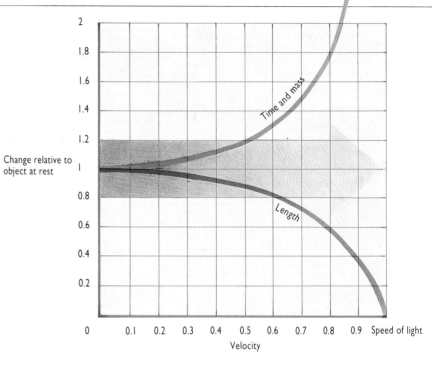

Before the work of Albert Einstein in the early 20th century, scientists assumed that time is the same everywhere in the universe – an hour is an hour, and time always progresses in a constant way. Einstein's Special Theory of Relativity shattered all that. He showed that time is entirely relative and passes faster or slower, depending on your position in space and the speed at which you are moving.

Scientists have known for a long time that motion is relative. Galileo, for instance, appreciated that a bottle sitting on a ship's table is motionless relative to the ship, but moves relative to the shore when the ship is moving.

Crucial to Einstein's insight was the discovery that there is an important exception to the fact that every motion is relative. This exception is light. The speed of light – some 186,300 miles/s (300,000 km/s) – is absolute. Nothing can travel faster than light, and light always travels at the same speed no matter where and how you measure it. So even if you travel at 99 percent of the speed of light, a beam of light will overtake you not just 1 percent faster, but the speed of light faster. Yet someone standing and watching you will see only 1 percent difference between your speed and that of the overtaking beam of light.

This seems illogical until it is appreciated that light is special. Photons, the "particles" of electromagnetic radiation, travel at the speed of light because they have no mass.

Relative speeds can be understood in everyday situations. For instance, if a motorcycle traveling at 125 mph (200 km/h) overtakes another going at 115 mph (185 km/h), the slower rider sees the faster machine going by at, effectively, 10 mph (15 km/h). But near the speed of light, relative speeds become paradoxical. If the slower bike could travel at just 10 mph (15 km/h) less than the speed of light, and the faster one at the speed of light, the slower bike would see the faster bike pass at the speed of light, not at 10 mph (15 km/h).

The speed of light limit has weird relativistic effects on objects moving at high speeds relative to observers. The three main effects are an increase in mass, a shrinking in length, and a slowing down of time (*left*). So an object traveling at 0.8 of the speed of light shrinks to about 0.6 of its length, is 1.8 times more massive, and its time passes 1.8 times more slowly.

As objects being observed get close to the speed of light, the effects increase dramatically. At the speed of light, an object's mass becomes infinite, its length diminishes to nothing, and time runs so slowly that it stops altogether. From the point of view of an object, however, it is the observers that appear to be moving, so the effects apply to them.

An object with mass cannot reach the speed of light because at the speed of light its mass becomes infinitely large. This means that an infinite force would be needed to drag it up to the speed of light, and there is no such thing as an infinite force.

Another relativity effect is that objects traveling near the speed of light appear to shrink lengthwise. At 90 percent of the speed of light, they seem to be less than half their original length. While fast-moving objects appear to contract from the standpoint of an outside observer, they do not actually shorten; the effect appears real nonetheless.

Length contraction leads to the third main relativity effect. If the length of a fast-moving object – a rocket, say – appears to contract, yet its speed is fixed, the time between the ticks of a clock aboard the rocket must appear to stretch, so time runs slower. At 90 percent of the speed of light, the clock takes over two "normal" seconds to tick off a second. Time stretching, or time dilation, is again only seen by an outside observer; anyone on board the rocket would see the outside observer's time stretch instead.

Communications satellite

Because light takes time to cover distance, we never see events at the exact time that they happen. When you look at your watch, for instance, you see the time fractionally after that shown on the watch, since light takes time to reach your eye. Similarly, when astronomers look at a galaxy 5 billion light-years away, they see it as it was 5 billion years ago, not as it is now. From elsewhere in the universe, it is seen at a different time. So the timing of events depends entirely on where you are.

Einstein started from this perspective and incorporated into his theories of relativity all the effects on time of the speed of light limit. He showed that the timing of events depends not only on where you are, but also on the speed at which time runs. He showed that there is no such thing as absolute time; time is only relative.

Few relativity effects play much of a part in our lives, but the time lag before hearing a reply when having a phone conversation with someone on the other side of the world (*left*) shows light's fixed speed. Calls are sent as radio signals via satellite. The lag is the time taken for a radio signal – traveling at the speed of light – to get up to the satellite, down to the other caller, and then back again. It can be nearly a second.

Sydney

Los Angeles

Curved space

Gravity may be nothing more than a twist in the fabric of space and time.

Although gravity is often described as a force that pulls matter together across space, the question of how this force is transmitted is unanswered. Most scientists believe that part of the solution lies in Einstein's General Theory of Relativity, published in 1916, which explained the significance of the acceleration produced by the force of gravity. Although not accepted by all scientists immediately after its publication, the theory has come into its own in recent decades, with the discovery of the way space curves, time dilation, and black holes.

Important in the General Theory is the notion of space-time. Conventionally, space is viewed only in the familiar three dimensions: height, width, and depth. But space and time are intimately bound, so space-time includes time as an extra dimension, and objects are thus located in

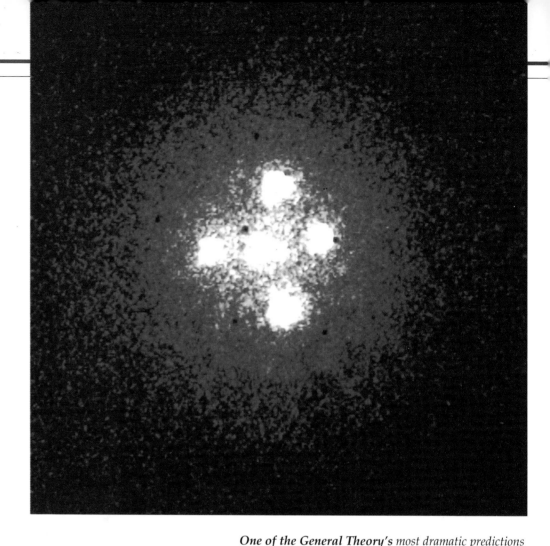

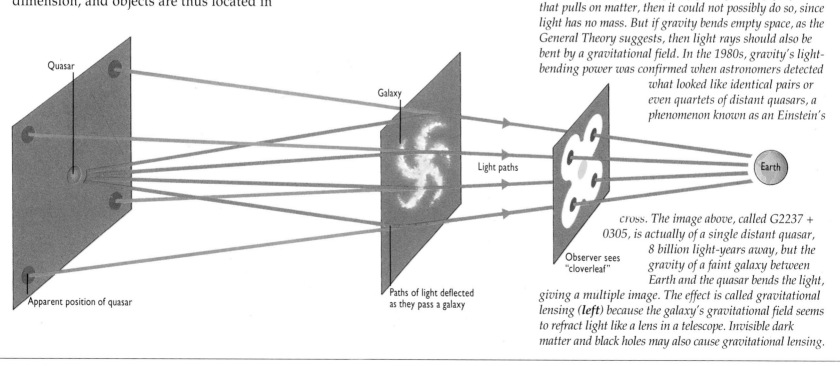

Quasar

Galaxy

Light paths

Earth

Apparent position of quasar

Paths of light deflected
as they pass a galaxy

Observer sees
"cloverleaf"

One of the General Theory's most dramatic predictions was that gravity bends light. If gravity is simply a force that pulls on matter, then it could not possibly do so, since light has no mass. But if gravity bends empty space, as the General Theory suggests, then light rays should also be bent by a gravitational field. In the 1980s, gravity's light-bending power was confirmed when astronomers detected what looked like identical pairs or even quartets of distant quasars, a phenomenon known as an Einstein's cross. The image above, called G2237 + 0305, is actually of a single distant quasar, 8 billion light-years away, but the gravity of a faint galaxy between Earth and the quasar bends the light, giving a multiple image. The effect is called gravitational lensing (left) because the galaxy's gravitational field seems to refract light like a lens in a telescope. Invisible dark matter and black holes may also cause gravitational lensing.

174

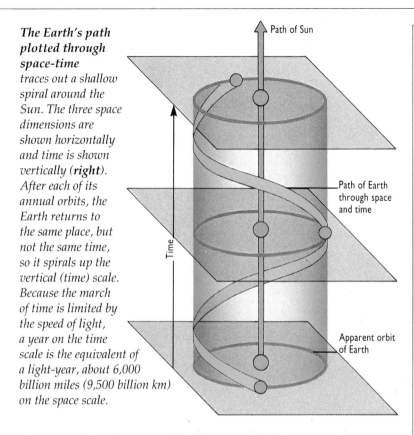

The Earth's path plotted through space-time traces out a shallow spiral around the Sun. The three space dimensions are shown horizontally and time is shown vertically (*right*). After each of its annual orbits, the Earth returns to the same place, but not the same time, so it spirals up the vertical (time) scale. Because the march of time is limited by the speed of light, a year on the time scale is the equivalent of a light-year, about 6,000 billion miles (9,500 billion km) on the space scale.

Path of Sun

Path of Earth through space and time

Time

Apparent orbit of Earth

time as well as in space. Tracing an object's space-time path therefore involves plotting its journey in both time and space.

Einstein's theory also shows that gravity works by bending space-time; matter simply follows the bend. If space-time is seen as a stretched rubber sheet, massive objects on it tug it down locally – the more massive the object, the bigger the distortion – and matter in the vicinity simply rolls into one of these dips. It is wrong, however, to think of gravity as a mysterious force pulling space-time into a curve; gravity is the curvature of space-time.

It is hard to imagine curved space. But what it means in practice is that gravity makes moving objects (and everything in space is moving) follow a curved path. However, it is important to remember that gravity is a curve in space-time, not just in space. The Earth's path through space is drawn into a curve by the gravity of the Sun. The curvature is not just Earth's elliptical three-dimensional orbit, but its spiral orbit through space-time. This is because the Earth returns to the same place again and again, but each year moves farther on in time.

ALBERT EINSTEIN

German-born Swiss physicist Albert Einstein (1879–1955) is the towering figure of 20th-century science. His ideas have created a revolution in the way we think about the universe. He is regarded as one of science's great geniuses for his original and challenging insights. In 1905 he published his Special Theory of Relativity, demolishing the notion of absolute time – which had prevailed since the age of Newton – and showing that all time is relative. It also reintroduced the idea of light as a particle, after a century's evidence had seemed to confirm that it was a wave. In 1916, he went further with his General Theory of Relativity, which showed how gravity, space, and time are inextricably linked in the universe.

In 1919 astronomers detected a slight shift in the apparent position of stars near the edge of the Sun's disk while monitoring a total eclipse (*below*) at Príncipe Island off the western coast of Africa. The shift not only showed that the Sun's gravity bends light rays from these distant stars, but also verified Einstein's General Theory.

175

Echoes of the Bang

It seems likely that the entire universe sprang into life at a single moment. How have astronomers come to this conclusion?

If you were looking for evidence of an explosion, the first thing you might investigate would be whether any matter was flung out and, if so, how it was moving today. But matter is all around, and it seems that in the universe as a whole, all matter is moving in a specific way.

For instance, looking at the way light from distant galaxies is redshifted reveals an astonishing fact – that all these galaxies are speeding away from the Earth. Indeed, the farther away a galaxy is, the greater its red shift and the faster it would seem to be disappearing into the void. This relationship is known as Hubble's Law, after the astronomer who discovered it in 1929. The most distant quasars are receding from us at speeds of up to 94 percent of the speed of light. Moreover, the farther away galaxies are, the farther apart they seem to be.

The only reasonable explanation for this is that the entire universe is expanding, and the galaxies are flying apart like the glowing embers from an exploding firework. But if the universe is continually expanding now, it must have been smaller in the past.

Putting the expansion into reverse and tracing it back in time leads to the inevitable conclusion that there was a moment – perhaps 15 billion years ago – when all the universe was compacted together as a single tiny volume. Most astronomers believe the universe burst into existence from this point with a Big Bang – an explosion of such power that the material of the universe is still hurtling away from it in all directions.

Although a few astronomers still dispute the Big Bang theory, there is one powerful piece of evidence in its favor: the echoes of this cataclysmic event can still be seen today.

*Galaxies today are racing apart in the wake of the Big Bang like the sparks from a firework (**right**). But it is wrong to think of the galaxies themselves moving through space. The most distant galaxies are not actually traveling at near the speed of light, even though they are moving away from us at this rate. It is rather the space in between galaxies that is stretching – and the huge red shifts observable from distant objects are the effect of the light waves stretching as they travel across the ever-widening void.*

In this COBE satellite microwave map of the sky, the purple areas are 0.01 percent warmer and the blue areas are 0.01 percent cooler than the 2.73K background temperature of the universe. These fluctuations hint at the variations in density needed for galaxies to form.

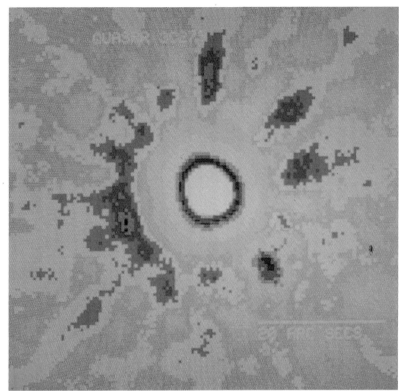

Evidence that the universe is expanding possibly comes also from distant quasars. Light left quasar 4C41.17, the most distant galaxy known, when the universe was barely one-fifth of its current age and is only reaching us now. The light is redshifted by an amount that indicates that it is moving away from us at 94 percent of the speed of light.

Faint traces of microwave radiation are coming from every direction in space. This cosmic background radiation is exactly the temperature (2.73K) that astronomers calculated it would be if it had cooled steadily since the Big Bang.

When the cosmic background radiation was first detected in the 1960s, it provided a major boost for the Big Bang theory. But for galaxies to have been able to form and come together in the time estimated to have passed since the Big Bang, there must have been hot spots in the early era of the universe – tiny density fluctuations which provided the seeds for galaxies to grow. Yet the cosmic background radiation seemed to be perfectly constant right across the sky, with no spots. In 1992, however, the COBE (Cosmic Background Explorer) satellite showed that there are very slight variations in the temperature of the background radiation. These 15-billion-year-old "cosmic ripples" detected by the satellite are the most convincing evidence yet of the Big Bang theory.

In the beginning

The universe came into existence about 15 billion years ago with the biggest explosion of all time – the Big Bang.

4 billion years
First galaxies appear

Time from Big Bang

Protogalaxies
and quasars

Expanding primordial gas

2 billion years
Gases cool and condense

Cosmologists have drawn up a detailed picture of how the universe came into being, back to within a mere 10^{-43} seconds of its beginning. This picture, known as the standard model of the Big Bang, starts with the universe incredibly hot and incredibly small, much smaller than an atom.

Suddenly, this ball began to expand at an astonishing rate, mushrooming to the size of our solar system within the first picosecond (10^{-12} seconds) and cooling rapidly as it did so, from 10^{32} K (ten thousand billion billion billion K) to 10^{16} K (ten million billion K). Within three minutes, the universe was many light-years across and the temperature had dropped to below 1 billion K, that is, to about 70 times hotter than the interior of the Sun today. As time went on, the rate of cooling dropped rapidly, and the expansion slowed down as gravity began to take effect.

At the temperatures of the early universe, matter and forces were quite unlike anything we know today. At the beginning, the four fundamental forces were melded into one, and the universe was filled with a broth of strange particles. Only as it began to cool did more familiar forces and particles begin to appear.

For the first few hundred thousand years, matter and radiation were thoroughly mixed up in a dense "fog." Some 300,000 years after the Bang, the temperature fell to 3,000K, electrons began to bind to hydrogen and helium nuclei to form the first stable atoms, and the fog cleared. Soon the universe was filled with swirling, primordial gas clouds that gradually curdled into long, thin strands separated by vast dark voids. These strands began to clump into galaxies, and 4 billion years later galaxies with stars existed.

Starting from the Bang, in the first 10^{-43} seconds, a time so short that it has its own name – Planck time – the universe was incredibly hot, at 10^{32} K, and the four forces were unified. As the temperature dropped to 10^{28} K, matter began to form in the shape of a swirling soup of quarks and leptons (electrons and neutrinos), and gravity split off from the other three forces. Between 10^{-35} and 10^{-32} seconds, the universe grew fast – the period of inflation – and as the temperature fell, the strong nuclear force separated out, leaving the electroweak force.

Meanwhile, matter and antimatter, the mirror image of matter, annihilated each other; there was just a little more matter than antimatter; this tiny surplus is what survives today. After 10^{-6} seconds, the temperature had dropped to 10^{13} K, and quarks began to combine while the weak nuclear and electromagnetic forces split. By 1 second, the temperature was 10^{10} K, and quarks had formed electrons and neutrinos and grouped together in triplets to make hadrons (protons and neutrons). At 3 minutes, the temperature was 10^9 K, and neutrons and protons began to stick together to form atomic nuclei such as deuterium and helium. After 10,000 years, the temperature dropped to 10^5 K and atoms began to form.

300,000 years
Stable atoms form; radiation and matter separate

3 mi

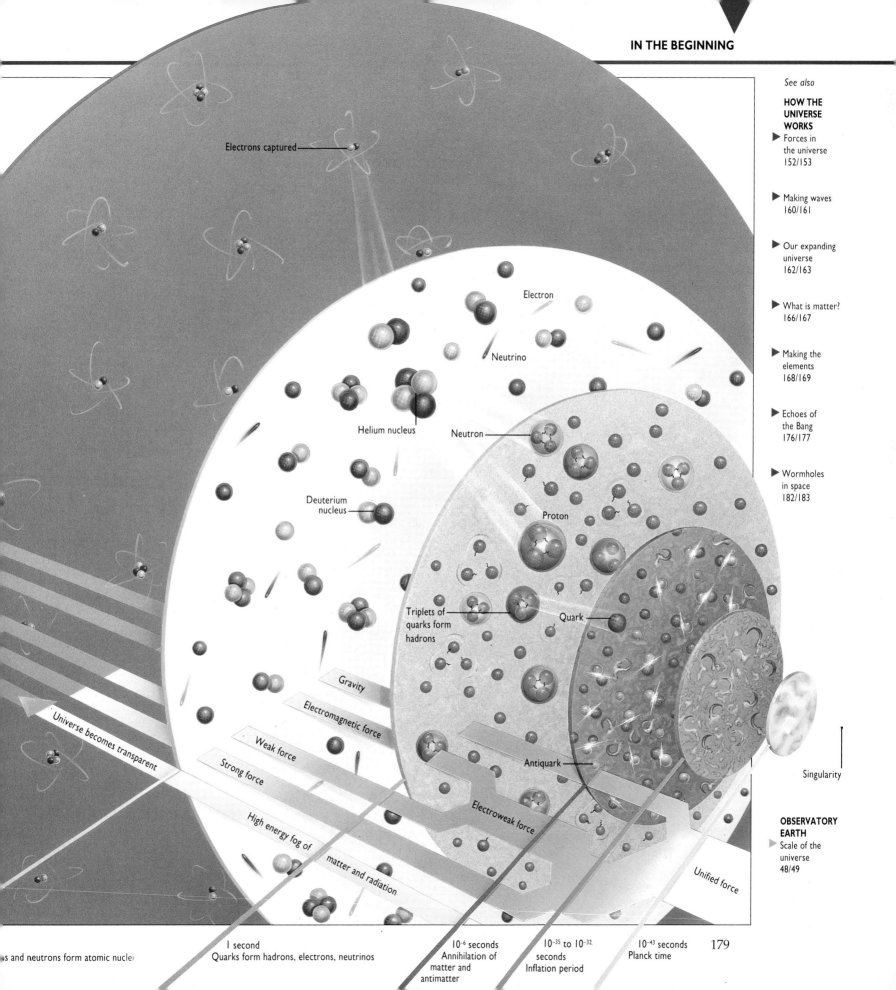

Electrons captured

Electron

Neutrino

Helium nucleus

Neutron

Deuterium
nucleus

Proton

Triplets of
quarks form
hadrons

Quark

Gravity

Electromagnetic force

Weak force

Strong force

Antiquark

Electroweak force

High energy fog of

Universe becomes transparent

matter and radiation

Unified force

Singularity

s and neutrons form atomic nuclei

1 second
Quarks form hadrons, electrons, neutrinos

10^{-6} seconds
Annihilation of
matter and
antimatter

10^{-35} to 10^{-32}
seconds
Inflation period

10^{-43} seconds
Planck time

179

The missing mass

The quest is on to learn the fate of the universe and to discover why much of its mass seems to be lost.

One of the most vexed questions facing astronomers is that of how much matter there is in the universe. Estimates of the amount are made by counting up the number of galaxies visible and multiplying the figure by their average mass, that is, the mass of the stars, gas, and dust that can be detected in them. Remarkably, the results suggest that there is barely 1 percent of the mass that the universe should contain according to the Big Bang theory. There must thus be a vast amount of "dark" matter that we simply cannot see. But although invisible, it should give itself away through its gravitational effects, acting, for instance, as a gravitational lens, distorting light from distant galaxies. In recent years, the search for missing mass has centered on identifying such lenses.

What is dark matter? If there was only a small amount more dark than visible matter, it could be ordinary matter, hidden perhaps in dark planets or in the dull embers of small stars, such as brown dwarfs, or even squashed into black holes. But few astronomers think that large quantities of dark matter are hidden in this way.

Some believe that dark matter lurks in the huge voids between galactic superclusters. Others hold that the mass exists in forms of matter unlike any yet discovered – so the hunt is on for new particles lurking in deep space.

*If the amount of matter in the universe hovers around the critical density – the amount needed for the gravitational force it creates to stop the universe from expanding – the galaxy would continue to expand for many billions of years to come. But gradually, its rate of growth would slow down until it reached a certain size, beyond which it would neither grow nor shrink. In this scenario the dots on the dice eventually reach a point where they remain the same distance apart for all time (**right, center**).*

*Whether the universe will continue to expand at its present astonishing rate depends on how much matter it contains and thus on how powerful a brake gravity puts on its expansion. If there was too little matter to slow expansion, matter and energy would eventually be spread so thinly that the universe would end up cold and empty, with immeasurably huge distances between galaxies. This scenario is known as the Big Chill (**below, top**). The galaxies would get farther and farther apart like the dots on the dice, forever receding.*

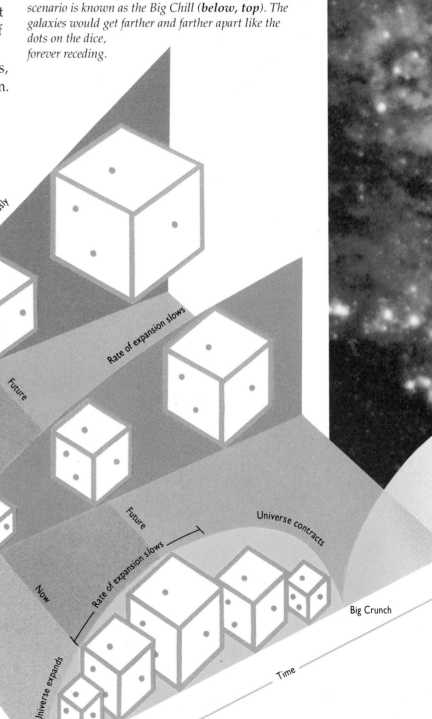

The mass of spiral galaxies, like M100, or NGC 4321, photographed here in great detail by the repaired Hubble Space Telescope in late 1993 (**left**), usually turns out to be much greater than expected, given the quantity of stars, dust, and gas visible. For instance, the orbits of globular clusters suggest that our galaxy's main disk is surrounded by a huge globe of dark matter. And, in the 1970s, astronomers studying the rotation of spiral galaxies discovered that, in every one they looked at, stars far from the center orbited almost as fast as those near the center. This implies that the gravitational field of a galaxy – and thus its mass – is about three times more than can be accounted for by the mutual attraction of its visible matter.

At about the same time, other astronomers, who were studying the gravitational stability of spiral galaxies, concluded that the spirals would fragment if they contained nothing but the matter we see. For the spiral to maintain its shape, the visible matter must be embedded in a supporting halo of dark matter – like the cream swirling on the top of a cup of coffee (**below**).

If there was a great deal more matter in the universe than there would currently appear to be, then the mutual attraction of all the mass would soon start to slow the expansion down. The dots on the dice would get a certain distance apart and then start to come closer together (**left, bottom**). The then-shrinking universe would accelerate inward. Eventually, the universe would end up in a reverse of the Big Bang – the Big Crunch – which might be followed by another Big Bang.

Wormholes in space

There may be a possibility that one day we might be able to travel backward and forward in time.

Ever since Einstein showed that time is relative and how gravity warps space-time, our notions of time and place have been fundamentally shaken. Einstein's Special Theory of Relativity revealed how time could slow down, and how someone journeying close to the speed of light might return home to find that a person born at exactly the same time had aged much faster. Then in the 1930s, American mathematician Kurt Gödel demonstrated how equations from Einstein's General Theory of Relativity opened up the theoretical possibility of traveling back in time.

But the real interest in time travel has been sparked by the idea of black holes, which are sites of the most extreme warping of space-time that we know of. Most theoretical pictures suggest that anything falling into a black hole is drawn inexorably toward the singularity at its center – that is, an infinitely small point where all time, space, matter, and energy end. But there are solutions to Einstein's equations which suggest that a traveler might avoid the

singularity and pass instead through a small passage to emerge at the far end through a "white hole" in another universe or another part of our universe. A white hole is the exact reverse of a black hole, and it is thought to be a place where matter and energy gush out like water from a fountain – just as in the Big Bang. Quasars and other explosions in the universe could be just such white holes. Most cosmologists feel that even if white holes exist, the weight of argument is firmly against any possibility of escaping the singularity. But could such passages exist without a singularity?

American astronomer and author Carl Sagan suggested the idea of wormholes through which a traveler might reach distant parts of the universe, or even different universes. The classical physics of Newton and Einstein indicates that a wormhole would pinch off, or slam shut, as soon as a traveler entered it. But quantum physics suggests there could be an escape. As a result, wormholes are now being investigated by astrophysicists at the California Institute of Technology.

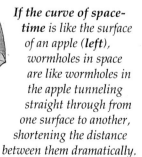

If the curve of space-time is like the surface of an apple (left), wormholes in space are like wormholes in the apple tunneling straight through from one surface to another, shortening the distance between them dramatically.

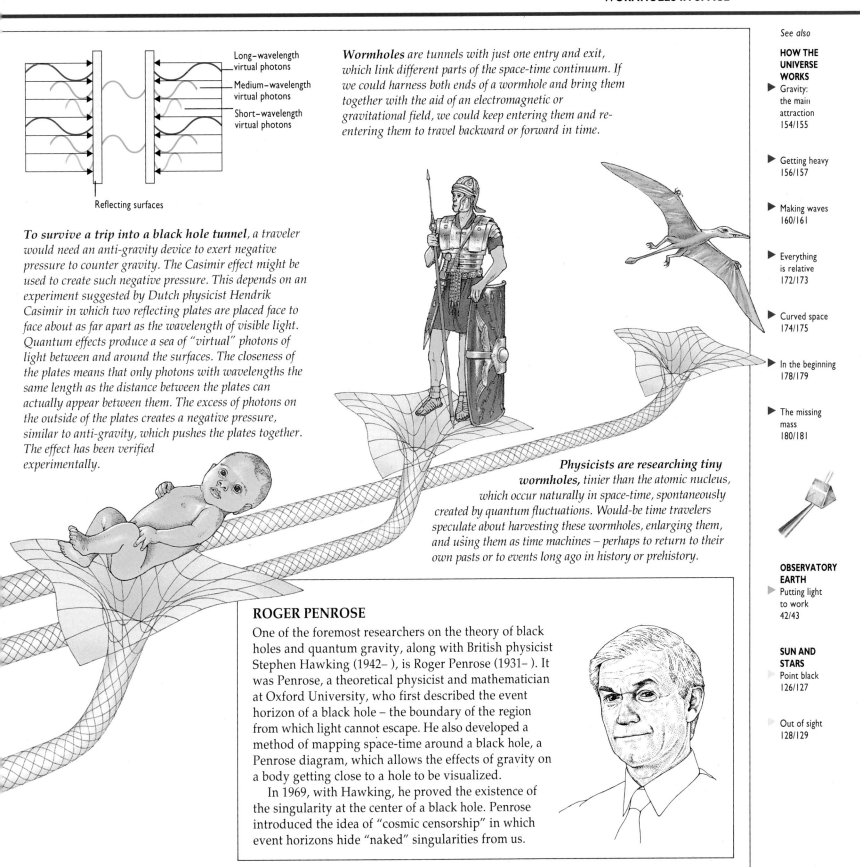

Long-wavelength virtual photons

Medium-wavelength virtual photons

Short-wavelength virtual photons

Reflecting surfaces

Wormholes are tunnels with just one entry and exit, which link different parts of the space-time continuum. If we could harness both ends of a wormhole and bring them together with the aid of an electromagnetic or gravitational field, we could keep entering them and re-entering them to travel backward or forward in time.

To survive a trip into a black hole tunnel, a traveler would need an anti-gravity device to exert negative pressure to counter gravity. The Casimir effect might be used to create such negative pressure. This depends on an experiment suggested by Dutch physicist Hendrik Casimir in which two reflecting plates are placed face to face about as far apart as the wavelength of visible light. Quantum effects produce a sea of "virtual" photons of light between and around the surfaces. The closeness of the plates means that only photons with wavelengths the same length as the distance between the plates can actually appear between them. The excess of photons on the outside of the plates creates a negative pressure, similar to anti-gravity, which pushes the plates together. The effect has been verified experimentally.

Physicists are researching tiny wormholes, tinier than the atomic nucleus, which occur naturally in space-time, spontaneously created by quantum fluctuations. Would-be time travelers speculate about harvesting these wormholes, enlarging them, and using them as time machines – perhaps to return to their own pasts or to events long ago in history or prehistory.

ROGER PENROSE

One of the foremost researchers on the theory of black holes and quantum gravity, along with British physicist Stephen Hawking (1942–), is Roger Penrose (1931–). It was Penrose, a theoretical physicist and mathematician at Oxford University, who first described the event horizon of a black hole – the boundary of the region from which light cannot escape. He also developed a method of mapping space-time around a black hole, a Penrose diagram, which allows the effects of gravity on a body getting close to a hole to be visualized.

In 1969, with Hawking, he proved the existence of the singularity at the center of a black hole. Penrose introduced the idea of "cosmic censorship" in which event horizons hide "naked" singularities from us.

183

We're here because we're here

So many factors had to be "just right" for life to come about, that it seems as though the universe must have been created specifically for this purpose.

Too cold to
sustain life

Mars

Is the emergence of life in the universe, especially human life, more than just a chance occurrence? Life as we know it depends on the presence of elements such as carbon, oxygen, nitrogen, and phosphorus. Yet only hydrogen and helium formed immediately after the Big Bang. Before the more complex elements could form, stars had to come into existence to forge them in their fiery interiors, then run their natural life to release them into space. To allow these elements to

evolve, the balance between three of the four forces – electromagnetism and the two nuclear forces – had to be just right. And for life to begin, these elements had to come together to form first complex amino acids and then self-reproducing chemicals in an environment in which life could be sustained. This extraordinary process has taken all the time in the universe – the entire 15 billion or so years since it formed.

The fact that all these apparently chance happenings have come about and allowed the development of intelligent life has led scientists to put forward the anthropic principle in its various versions. According to the weak anthropic principle, only a universe with an appropriate construction is likely to contain intelligent life. Some scientists go further and propose the strong anthropic principle. In this, the universe had to bring intelligent life into existence at some point in its history and so must have those properties that allow such life to develop. The final anthropic principle states that since intelligent life has appeared, it will never die out.

The first life forms were probably single-celled bacteria consisting of an envelope of chemicals including vital DNA – the carrier of genetic information. From these evolved cells with areas for specialist functions, each with a nucleus holding the DNA. Multi-function cells evolved into the animal cells from which we are all constructed.

Complex space missions, such as the repair of the Hubble Space Telescope late in 1993 (above), show off our intelligence. But are we the only intelligent species? The universe is so vast that surely we are not alone. Somewhere else the conditions must have arisen for intelligent life to evolve.

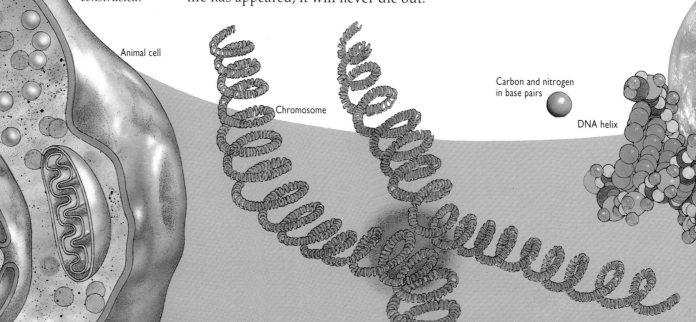

Animal cell

Chromosome

Carbon and nitrogen
in base pairs

DNA helix

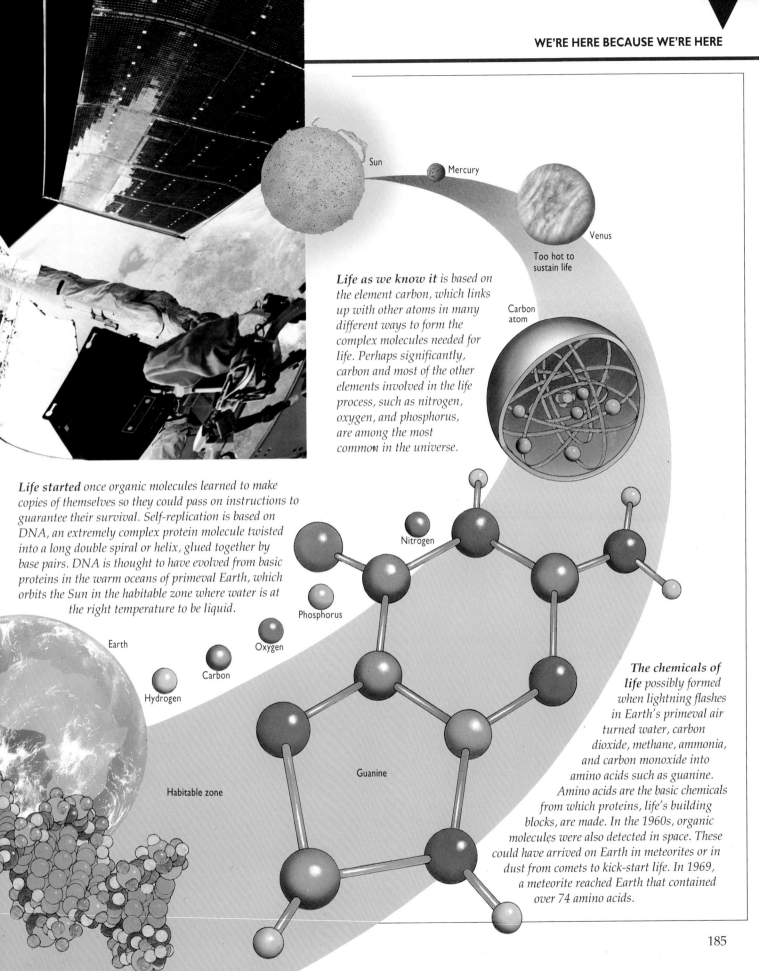

Sun

Mercury

Venus

Too hot to
sustain life

Carbon
atom

Life as we know it is based on
the element carbon, which links
up with other atoms in many
different ways to form the
complex molecules needed for
life. Perhaps significantly,
carbon and most of the other
elements involved in the life
process, such as nitrogen,
oxygen, and phosphorus,
are among the most
common in the universe.

Life started once organic molecules learned to make
copies of themselves so they could pass on instructions to
guarantee their survival. Self-replication is based on
DNA, an extremely complex protein molecule twisted
into a long double spiral or helix, glued together by
base pairs. DNA is thought to have evolved from basic
proteins in the warm oceans of primeval Earth, which
orbits the Sun in the habitable zone where water is at
the right temperature to be liquid.

Nitrogen

Phosphorus

Earth

Oxygen

Carbon

Hydrogen

Habitable zone

Guanine

*The chemicals of
life* possibly formed
when lightning flashes
in Earth's primeval air
turned water, carbon
dioxide, methane, ammonia,
and carbon monoxide into
amino acids such as guanine.
Amino acids are the basic chemicals
from which proteins, life's building
blocks, are made. In the 1960s, organic
molecules were also detected in space. These
could have arrived on Earth in meteorites or in
dust from comets to kick-start life. In 1969,
a meteorite reached Earth that contained
over 74 amino acids.

185

Bibliography

Audouze, Jean and Guy Israël (eds.) *The Cambridge Atlas of Astronomy* Cambridge University Press, Cambridge, 1988

Beatty, J. Kelly and Andrew Chaikin (eds.) *The New Solar System* Sky Publishing Corporation, Cambridge, Mass., 1990

Calder, Nigel *Spaceship Earth* Viking, London, 1991

Cattermole, Peter, Garry Hunt, Patrick Moore and Iain Nicolson *The Atlas of the Solar System* Mitchell Beazley, London, 1983

Chown, Marcus *Afterglow of Creation* Arrow Books, London, 1993

Close, Frank, Michael Marten and Christine Sutton *The Particle Explosion* Oxford University Press, Oxford, 1987

Couper, Heather and Nigel Henbest *The Space Atlas* Dorling Kindersley, London, 1992

Coveney, Peter and Roger Highfield *The Arrow of Time,* Flamingo, London, 1993

Cox, P.A. *The Elements: Their Origin, Abundance and Distribution* Oxford University Press, Oxford, 1989

Davies, Paul (ed.) *The New Physics* Cambridge University Press, Cambridge, 1989
——*Superforce* Unwin, London, 1985

Davies, Paul and John Gribbin *The Matter Myth* Viking, London, 1991

de la Cotardiere, Philippe *Larousse Astronomy: New Revised Edition* Hamlyn, London, 1987

Dunlop, Storm and Colin Ronan (eds.) *The Skywatchers Handbook* Crown Publishers, New York, 1989

Galaxies (Voyage Through the Universe Series) Time-Life Books, Amsterdam, 1988

Gribbin, John *In the Beginning* Viking, London, 1993
——*In Search of Big Bang* Corgi, London, 1987

Hawking, Stephen *A Brief History of Time* Bantam, New York, 1988

Henbest, Nigel and Michael Marten (eds.) *The New Astronomy* Cambridge University Press, Cambridge, 1985

Hunt, Garry and Patrick Moore *Atlas of Uranus* Cambridge University Press, Cambridge, 1989

Kerrod, Robin *The Illustrated History of Man in Space* Prion, London, 1989
——*The Heavens: Planets, Stars, Galaxies* The Leisure Circle Ltd., London, 1984

Lederman, Leon and David Schramm *From Quarks to the Cosmos* Scientific American Library, New York, 1989

Levy, David H. *The Sky: A User's Guide* Cambridge University Press, Cambridge, 1993

Luminet, Jean-Pierre *Black Holes* Cambridge University Press, Cambridge, 1993

Malin, David and Paul Murdin *Colours of the Stars* Cambridge University Press, Cambridge, 1984

Mitton, Jacqueline and Simon *The Prentice-Hall Concise Book of Astronomy* Prentice-Hall Inc., Englewood Cliffs, New Jersey, 1979

Moore, Patrick *The Guinness Book of Astronomy Facts and Feats* Guinness Superlatives Ltd., London, 1983
——*The Mitchell Beazley Concise Atlas of the Universe* Mitchell Beazley, London, 1974

Moore, Patrick and Iain Nicolson (eds.) *The Universe* Collins, London, 1985

Morrison, Philip and Phyllis *Powers of Ten* Scientific American Library, New York, 1982

The Near Planets (Voyage Through the Universe Series) Time-Life Books, Amsterdam, 1989

Penrose, Roger *The Emperor's New Mind* Oxford University Press, Oxford, 1989

Ridpath, Ian *Norton's 2000.0* Longman, Harlow, 1989
——*Longman Illustrated Dictionary of Astronomy and Astronautics* Longman, Harlow, 1987

Riordan, Michael and David Schramm *Dark Matter and the Structure of the Universe* Oxford University Press, Oxford, 1993

Ronan, Colin *Science Explained* Henry Holt, New York, 1993
——*The Natural History of the Universe* Macmillan, New York, 1991
——*Deep Space* Pan, London, 1983
——*The Practical Astronomer* Pan, London, 1981

Room, Adrian *Dictionary of Astronomical Names* Routledge, London and New York, 1988

Scagell, Robin *Astronomy from Towns and Suburbs* George Philip, London, 1994

Stott, Carole *The Greenwich Guide to Astronomy in Action* George Philip, London, 1989

Tirion, Wil *Sky Atlas 2000.0* Cambridge University Press, Cambridge, 1982

Trux, Jon *The Space Race* New English Library, Sevenoaks and London, 1985

Suggested journals and periodicals

Nature Macmillan Magazines, London
New Scientist IPC Magazines Ltd., London
Scientific American Scientific American Inc., New York
Sky and Telescope Sky Publishing Corporation, Cambridge, Mass.

A MATTER OF DEGREE

For everyday purposes, temperatures are measured in degrees on the Fahrenheit or Celsius scales. Both scales rely on fixed points, such as the boiling and freezing points of water, and simply divide intervening temperatures equally into degrees. Water boils at 212° on the Fahrenheit scale and 100° on the Celsius scale; it freezes at 32° on the Fahrenheit scale and 0° on the Celsius scale.

Most scientists, however, use the Kelvin, or absolute, scale. On this scale, 0K is absolute zero, the coldest temperature theoretically possible. At this temperature, all molecules would stop moving. 0K is actually −459.67°F (−273.15°C). The temperature intervals of the Kelvin and Celsius scales are equal. So 0°C is 273.15K and 100°C is 373.15K. Extremes of temperature can be created in laboratories, from only a millionth of a degree above absolute zero to the billions of degrees that the universe experienced within a second of the Big Bang.

HIGHER POWERS

Very large numbers and fractions can be written compactly in multiples of ten. Thus 100 is 10^2 since it is 10 x 10, or ten to the power two; 1,000 is 10^3 or ten to the power three; a million is 10^6, a billion is 10^9. Using this method, a number such as 3,600,000 can be written as 3.6×10^6 ($3.6 \times 1,000,000$).

Numbers less than 1 can also be written in this compact way. For example, 10^{-3} represents 1 divided by 10^3, or one-thousandth (0.001), so 3.6×10^{-6} represents 0.0000036.

Index

Page numbers in *italic* refer to illustrations and box text.

Acknowledgments

l=left; *r*=right; *c*=center; *t*=top; *b*=bottom

Picture credits
2 NASA; 6*tl* AT&T; 6*tr* John Barlow; 6*c* NASA; 6*b* John Barlow; 10*l* John Barlow; 10*tr* British Library; 10*br* Ian Morison/Jodrell Bank; 11 Sonia Halliday Photographs; 12 David Malin/Anglo-Australian Observatory; 12/13 NASA; 13 JPL/NASA; 14 Robin Scagell/Galaxy Picture Library; 14/15 Dennis di Cicco/Sky Publishing; 18*l* Meade Instruments; 18*r* John Barlow; 19*t* John Barlow; 19*b* Meade Instruments; 20*l* R. Garner/Galaxy Picture Library; 20*r* Alex Colburn/Galaxy Picture Library; 21*t* Roger Ressmeyer, Starlight/Science Photo Library; 21*b* NOAO; 22 British Library; 30 Lick Observatory; 30/31 Robin Scagell/Galaxy Picture Library; 32 John Barlow; 32/33 Michael Maunder/Galaxy Picture Library; 33 Paul Doherty; 34/35 J. & D. Begg/Ace Photo Agency; 36/37 D.A. Calvert/Royal Greenwich Observatory; 37 D.G. Sythes/Whitby Museum; 40/41 Andy Caulfield/The Image Bank; 42*l* A. Pasieka/The Image Bank; 42*r* Dept. of Physics, Imperial College/Science Photo Library; 43*tl* IDSA/Galaxy Picture Library; 43*tr* Sonia Halliday Photographs; 43*bl* Robin Bath; 43*br* Dept. of Physics, Imperial College/Science Photo Library; 44/45 NASA; 45*t* Dr. Eli Brinks/Science Photo Library; 45*b* Ian Morison/Jodrell Bank; 46/47*tl* NASA; 47 David Malin/Anglo-Australian Observatory; 50*t* NASA; 50*b* John Barlow; 51 USGS/NASA; 53*t* Royal Astronomical Society Library; 53*b* Science Graphics Inc.; 56/57 NASA; 57 John Barlow; 60/61 Picturepoint; 62/63 Dr. David Gorham & Dr. Ian Hutchings/Science Photo Library; 63 Lick Observatory; 64*t* D.W. Wilhelms/Academic Press; 64*b* Apollo 16 Principal Investigator/Frederick J. Doyle/NSSDC; 65 D.W. Wilhelms/Academic Press; 66/68 NASA; 69 Royal Greenwich Observatory; 70/71*t* NASA; 71*c* Martyn Chillmaid/Oxford Scientific Films; 71*bl* USGS/NASA; 71*br* JPL/NASA; 72*t* John Sanford/Science Photo Library; 72*b* Royal Astronomical Society Library; 72/73 NASA; 74*t* Science Photo Library; 74*c* John Barlow; 74*b* NASA; 75*t* Obayashi Corporation; 75*b* NASA; 76/77*t* NASA; 77*b* John Barlow; 78/79 NASA; 79 Royal Society; 80/81 NASA; 82/84 NASA; 86/87 JPL/NASA; 88/89*t* JPL/NASA; 89*b* John Barlow; 90 NASA/ESA; 93 NASA; 94 Richard Fowell/Science Photo Library; 95 Barney Magrath/Science Photo Library; 96 NASA; 96/97 Pekka Parvianen/Science Photo Library; 97 François Gohier/Science Picture Library; 98 Brett Froomer/The Image Bank; 100*t* David Malin/Anglo-Australian Observatory; 100*b* Steve Parker/Ace Photo Agency; 101 The Image Bank; 102 John Barlow; 103 Dr. Kaler/The Astronomical Society of the Pacific; 104 NASA; 104/105 Dr. S. Koutchmy/CNRS; 105*t* John Barlow; 105*b* Royal Greenwich Observatory; 106 US Navy/Science Photo Library; 106/107 JET; 108 Hozelock Limited; 108/109*t* Pekka Parvianen/Science Photo Library; 108/109*b* JPL/NASA; 111 NASA; 112 Steve Parker/Ace Photo Agency; 114 Photo Library International/Science Photo Library; 114/115 Anglo-Australian Observatory; 116 Adrian Murrell/Allsport; 117 David Malin/Anglo-Australian Observatory; 118 Doug Plummer/Science Photo Library; 120 Lori Adamski Peek/Tony Stone Images; 121 NASA; 124/125 Anglo-Australian Observatory; 126/127 The Image Bank; 127 Max-Planck-Institut für Extraterrestrische Physik/Science Photo Library; 130*t* Anglo-Australian Observatory; 130*bl* John Barlow; 130*br*/131 David Malin/Royal Observatory Edinburgh; 132/133 Anglo-Australian Observatory; 134 John Barlow; 134/135 David Malin/Anglo-Australian Observatory; 136/137*l* David Malin/Royal Observatory Edinburgh; 137*r*/139 David Malin/Anglo-Australian Observatory; 140 Royal Observatory Edinburgh; 140/141 Zefa Picture Library; 142*l* NRAO/AUI; 142*c* Dr. Eric Feigelson/Science Photo Library; 142*r* David Malin/Anglo-Australian Observatory; 144 NASA; 144/145 National Optical Astronomy Observatories; 145 Lund Observatory; 147 David Malin/Royal Observatory Edinburgh/Anglo-Australian Observatory; 148/149 M. Seldner, B.L. Siebers, E.J. Groth and P.J.E. Peebles/Astronomical Journal 82, 249, 1977; 150*tl* John Barlow; 150*tr* Images Colour Library; 150*bl* Don Morley; 150*br* John Barlow; 152/153 F. Streichan/Zefa Picture Library; 154/155 Philippe Blondel/Allsport; 156/157 NASA; 157 The University of Western Australia; 158 John Barlow; 158/159 NASA; 160/161 Images Colour Library; 162*t* Don Morley; 162*b* John Barlow; 162/163 Don Morley; 164/165 John Barlow; 166/167 David Malin/Anglo-Australian Observatory; 169/170 John Barlow; 171*t* NASA; 171*b* CERN Photo; 172/173 Howard Boylan/Allsport; 174 Goddard Space Flight Center; 175 Royal Greenwich Observatory; 176 NASA; 176/177 Simon Wilkinson/The Image Bank; 177 Leicester University X-Ray Astronomy Group/Science Photo Library; 180/181 NASA; 181 John Barlow; 184/185 NASA

If the publishers have unwittingly infringed copyright in any illustration reproduced, they would pay an appropriate fee on being satisfied to the owner's title.

Illustration credits
David Ashby (portraits) 36, 45, 53, 85, 87, 103, 139, 144, 163, 175; Bill Donohoe 22/23, 67, 84/85, 112/113, 142/143, 182/183, 184/185; Andrew Farmer 38/39, 42, 46/47, 145*br*, 147*cr*, 148/149 (Earth time insets), 152/153, also connection icons; Chris Forsey 14/15, 30/31, 34/35, 43, 58/59, 70*tr*, 79, 82/83, 89, 92/93, 104, 118/119, 154/155, 156/157, 164, 172/173; Aziz Kahn 165; Line & Line 61; Mainline Design 16/17, 18/19, 24/25, 26/27, 28/29, 37, 46/47, 52/53, 54/55, 56, 62, 68/69, 70*cl*, 72, 76, 80/81, 82*cl*, 86, 88, 102/103, 110/111, 115, 116/117, 124, 126*bl*, 128/129, 138, 146/147, 148/149, 152*bl*, 158, 160/161, 166/167, 168/169, 170/171, 178/179; David Parker 18/19; Colin Salmon 13, 33, 60, 106/107, 108/109, 126, 180, 174/175; Mike Saunders 48/49, 65, 90/91, 98/99, 120/121, 122/123, 133, 135, 140/141, 145*c*, 145*tr*; Ed Stuart 73

Marshall Editions would like to thank the following for their assistance in the compilation of this book:

Authors: Nigel Cawthorne 52–91, 102–29
           John Farndon 132–49, 152–85
           Jon Kirkwood 11, 24–29, 51, 101, 131, 151
           Robin Scagell 12–23, 30–49, 92–99

Copy editors: Isabella Raeburn, Maggi McCormick
Managing editor: Lindsay McTeague
Editorial director: Ruth Binney
DTP editor: Mary Pickles
DTP assistance: Pennie Jelliff, Heather Magrill
Index: Caroline S. Sheard
Assistant art editor: Lynn Bowers
Picture manager: Zilda Tandy
Production: Janice Storr, Sarah Hinks

Broadhurst, Clarkson & Fuller Ltd. for the loan of astronomical equipment; Olympus Sport for the loan of clothes; Freewheel for the loan of bicycle equipment; Franco and Romano of Buona Sera Rosticceria for use of their facilities.